Spills and Spin

Tom Bergin has reported on the energy industry for over twelve years, having previously worked as an oil broker. For the past seven years he has headed Reuters's coverage of the oil industry in Europe, the Middle East and Africa, and his work has been published in *The New York Times*, *The Times*, *The Wall Street Journal*, *International Herald Tribune*, *The Globe and Mail* and the *Shanghai Daily*, as well as in dozens of newspapers and magazines around the world. He is a regular television and radio commentator, appearing on CNBC, ITV, the BBC and other outlets as far away as New Zealand.

Apart from the oil industry, Tom has reported on financial scandals, including the rise and fall of Enron, environmental issues, EU politics and terrorist attacks. He lives in London with his wife, a former Reuters reporter turned investment banker, and two young sons.

Praise for *Spills and Spin*

'Compelling . . . Bergin's account stands out as the most thoughtful of the many recent accounts'
Daily Telegraph

'Bergin, a highly regarded oil industry reporter for Reuters, has provided the best assessment yet of how the accident was rooted in the nature of BP . . . There are lessons here about how to prevent a crisis – and handle one – that anyone in business would do well to take in'
Financial Times

'Exhaustively researched . . . If you want to know why BP got itself into such a mess before and after Deepwater Horizon, *Spills and Spin* is an excellent and reliable account'
Sunday Times

'Here is the story not only of BP's darkest day, but of its 20-year transformation from takeover target to dynamic market leader . . . For a robust read on the combustible history of big oil . . . Bergin has all the raw material you require'
The Richmond Magazine

'Reveals how the Macondo well disaster was rooted to the history and culture of BP'
The Times

'Written in tremendous style . . . the passages explaining what went wrong on Deepwater Horizon are clear and balanced, though obviously still very controversial. He is also excellent at drawing out the consequences of BP's failures . . . [a] very persuasive account'
Financial World

TOM BERGIN

Spills and Spin

The Inside Story of BP

Published by Random House Business Books 2012

2 4 6 8 10 9 7 5 3

First published in Great Britain in 2011 by
Random House Business Books
Random House, 20 Vauxhall Bridge Road,
London SW1V 2SA

www.randomhouse.co.uk

Addresses for companies within The Random House Group Limited can be found at: www.randomhouse.co.uk/offices.htm

The Random House Group Limited Reg. No. 954009

A CIP catalogue record for this book is available from the British Library

ISBN 9781847940827

The Random House Group Limited supports The Forest Stewardship Council (FSC®), the leading international forest certification organisation. Our books carrying the FSC label are printed on FSC® certified paper. FSC is the only forest certification scheme endorsed by the leading environmental organisations, including Greenpeace. Our paper procurement policy can be found at:
www.randomhouse.co.uk/environment

Printed in Great Britain by Clays Ltd, St Ives plc

Contents

Introduction

Shortly before 10 p.m. on 20 April 2010, Captain Alwin J. Landry was sitting at his desk in the wheelhouse of the *Damon B. Bankston*, writing his log and catching up on paperwork. With three levels of cabins at the front and a long, low back, the 260-foot *Bankston* looked like a floating flatbed truck, a description that pretty much summed up its role: the offshore supply vessel was under charter to London-based oil giant BP, and spent its time ferrying equipment and supplies from Port Fourchon, Louisiana to the company's rigs in the Gulf of Mexico.

That night, the *Bankston* was moored alongside the *Deepwater Horizon* rig, linked to it via a heavy industrial hose. At any moment, Captain Landry was expecting *Horizon* to begin pumping drilling mud – a heavy fluid used to lubricate the drilling process – into tanks within the ship's hull. *Horizon* had just completed BP's 'Macondo' exploration well, and the removal of mud from the well was one of the last routine procedures it had to undertake before being moved on to its next job.

Suddenly first mate Paul Erickson appeared, asking the captain to come and inspect something that had unsettled him. Landry stared out of the window and saw the deck of his ship glistening in the moonlight: the entire back of the vessel was covered in drilling mud. At first he thought the hose was leaking – he had seen this happen before – but then he noticed the mud was falling from the sky like black rain. This he had not seen before. He went over to the port side of the wheelhouse, which was level with *Deepwater Horizon*'s upper deck, looked out and saw that the mud was emerging, like a fountain, from the top of the rig's derrick. A loud hissing noise indicated a gas release. As though to emphasise the sense of impending doom, a couple of sea birds fell dead to the deck, their feathers drenched in drilling mud. Landry picked up the radio to *Deepwater Horizon* and told them, 'I'm getting mud on me.'

By this stage, Curt Kuchta, captain of *Deepwater Horizon*, had also noticed something was wrong. From a window on the bridge of the rig he could see that there was mud on the water. Another window revealed mud coming out of a piece of equipment beside the derrick. He too could hear the hissing sound, although it was soon drowned out by the even more worrying noise of the rig's gas alarms.

Kuchta then overheard a colleague on the radio to Captain Landry, explaining that *Deepwater Horizon* was experiencing a 'well control incident'. He grabbed the microphone and ordered Landry to pull away from the rig immediately. Landry replied that he couldn't: the *Bankston* was still connected by the mud transfer hose. Kuchta paused for a moment, trying to decide what he should tell Landry to do. Before he could come up with an answer, a massive blast ripped through the main deck of the rig.

From the *Bankston*, Landry saw a green flash, a fireball and a wave of debris flying toward him. Only 40 feet from the rig, he could also feel the shock wave. Beside him, Paul Erickson began to shout, 'Fire! Fire! Fire on the rig!' and made for the alarm. Landry picked up the internal radio and ordered his men below to disconnect the mud hose immediately.

Landry revved the engines and pushed back. When he got to a distance of 100 metres, he saw *Deepwater Horizon* go dark. It had lost power, although he could still hear its distress signals over the radio. The *Bankston*'s satellite phone wasn't working – coverage was unreliable in this area – and so Landry dispatched an email to BP alerting them to the explosion. Meanwhile, secondary explosions began to rock the *Horizon*, turning the rig into one large fireball. Landry saw workers jump into the sea to escape the flames.

Below deck on the *Bankston*, engine room assistant Louis Longlois and chief engineer Anthony Gervasio were already preparing the fast recovery craft. Once the vessel had reached its 500-metre holding point, Landry ordered the small craft into the water to pick up the men floundering in the Gulf.

Over the coming hours, 115 men were recovered from lifeboats and life rafts and plucked from the burning water around the rig. It was a heroic achievement on the part of the *Bankston*'s crew, but the head-count was 11 fewer than had been on the rig before the blast.

Deepwater Horizon burned for another two days before sinking 5,000 feet to the bottom of the Gulf of Mexico, in the process unleashing the worst oil spill in US history.

Days later, when my colleagues and I at Reuters bureaux in London, Houston, Miami, New York and Washington held conference calls to plan our coverage of what was rapidly emerging as the biggest news story of 2010, there was a distinct sense of déjà vu.

BP was no stranger to disasters. Over the previous five years, it had been responsible for two of the highest-profile industrial accidents in the US: an explosion at Texas City and a series of oil spills in Alaska. The company's commercial activities had also been the source of relatively frequent controversy, sparking numerous US Congressional investigations over the course of 20 years. We all felt we had been here before.

We knew the public relations people BP had hired to help handle the media storm, the trial lawyers who would be leading the inevitable class action suits, and the staff of the Congressional committees that had immediately announced their intention to investigate the spill.

There was one key difference this time: the man at the top. BP's chief executive Tony Hayward had been credited with turning the company around over the three years of his tenure, following a number of disasters under his predecessor Lord John Browne. This was Hayward's first big challenge.

A week after the blast I had a conference call with him, and he sounded almost relaxed. A few days later, he came to my office in London for lunch, and although he could now see that the spill would be an event of major significance for the oil industry, he seemed surprisingly optimistic. He had an upbeat view of the financial impact on BP and of the environmental consequences for the Gulf.

In the event, the spill pushed BP to the brink of bankruptcy, partly due to Hayward's infamous PR gaffes, but more importantly because the American public, media and many politicians concluded that the disaster was yet another symptom of BP's culture of compulsive cost-cutting. This narrative wasn't necessarily wrong, but it didn't tell the full story. Low-fare airlines and pile-'em-high supermarket chains would be nothing without constant cost-cutting – yet their planes don't fall out of the skies and their food doesn't poison customers. The story of the oil spill was more complicated than that, not least because it came down to people, and ultimately hinged on two men in particular: John Browne, perhaps the unlikeliest of oil men, and Tony Hayward, a geologist from humble origins with a fierce ambition. The story of the oil spill is a 20-year tale of Browne's determination to create the largest corporation in the world, and Hayward's battle to succeed, and then outdo, his mentor.

This is that story.

1

The Sun King Rises

BP's 2009 annual general meeting in London was an even more colourful affair than usual. The proceedings were enlivened by the usual turnout of environmental activists and eccentric small shareholders.

The green groups had purchased a few shares in the company to give themselves the right to turn up and harangue the management over their environmental crimes. One representative reminded everyone present that BP's business model of producing hydrocarbons risked a climate catastrophe. A Native American dressed in jeans and a T-shirt argued that drilling for shale gas, something BP and a hundred other companies across the US practised, risked polluting groundwater.

The small shareholders were largely pensioners who saw the event as a day out. Some stood up and complained to the chairman about the choice of location, the ExCeL Centre, a large hangar-like convention hall in the Docklands, east of the City. It was cold and noisy, they said, and not nearly as convenient as the Royal Festival Hall in the Southbank complex, the location for previous AGMs.

The man who conducted forensic analyses of the annual report was there as usual, complaining about inadequate disclosure somewhere in this year's edition. The retired captain from the shipping division, who wore a long, bushy white beard, was there again, protesting the report's sparse mention of the shipping division. Another familiar face rose to enquire why the management were being paid bonuses. Surely, if they were motivated people, they didn't need bonuses to motivate them – and if they weren't motivated, they shouldn't be in their jobs.

Between them, the people gathered in the hall owned less than 1 per cent of the company's shares. The event was a formality and, after years in the job, BP's chairman Peter Sutherland had mastered his technique for handling the crowd.

Sutherland was a former international diplomat, previously a European Union Commissioner and head of the World Trade Organization. His approach to managing the AGM was a mixture of statesmanship and stand-up, deftly diverting the speakers with either respectful thanks or sardonic put-downs, and sometimes both. The man who complained that Sutherland had ignored his correspondence received an apology but was told that the sheer regularity, and length, at which he wrote meant that replying to each of the missives would preclude Sutherland from conducting any of his other duties as chairman.

Yet the proceedings were imbued with an extra sense of theatre this year. This was BP's centenary year: 100 years since the company's antecedent, the Anglo-Persian Oil Company, was floated on the stock exchange.

Next door to the auditorium, in another equally cavernous hall where investors would later be served boxed lunches and small plastic bottles of wine, BP had arranged an exhibition outlining the company's illustrious history. An ancient well head painted glossy black stood at the centre of a display that also contained antique geological instruments and handwritten journals from the original explorers who found oil in Persia.

Given the recent collapse of oil prices from almost $150 a barrel to under $40, chief executive Tony Hayward had ordered that the centenary celebrations be a low-key affair: no extravagant events and no fireworks or any of the razzmatazz favoured by his predecessor, Lord John Browne. It was a reflection of a cost-conscious approach to business that Hayward had intensified in the past year or two. After oil prices had tumbled, other oil companies had also started to re-examine their cost bases, but BP, Sutherland told the meeting, had 'been ahead of the game in addressing costs'.

One of the investors said he knew it: his company, a supplier of cement for lining oil wells, was feeling the pinch. Hayward said he wouldn't apologise for this and vowed to continue squeezing suppliers on the shareholders' behalf. The comments won applause. Hayward, investors agreed, was turning the business around after some tough years.

Of course, BP was no stranger to tough times, nor to overcoming them. Founded on the back of massive discoveries in Iran (or Persia, as it was then known), the Anglo-Persian Oil Company was floundering within a few years of listing on the stock market after it experienced difficulty in securing markets for its crude. The company was bailed out in 1914 by the British government taking a majority stake. Winston Churchill, then First Lord of the Admiralty, was instrumental in the decision. He believed control of the company would ensure a secure fuel supply for the Royal Navy, at a time when it was shifting from burning coal to oil.

In 1953, Churchill once again saved what was now called the Anglo-Iranian Oil Company, after Iran's democratically elected prime minister, Mohammed Mossadegh, nationalised the country's oil industry. Churchill encouraged the CIA, with help from MI6, to support a coup d'état to remove Mossadegh. They did just that and the company – soon renamed once again, as British Petroleum – got its oil fields back. With a foothold in the Middle East, BP branched out into other countries, including Iraq, where it found further huge reserves, and it was pumping 4.7 million barrels per day by the early 1970s. But the spirit of Mossadegh was being revived across the Middle East: nationalisations in Iraq, Libya and Abu Dhabi, and finally the deposition of the Shah of Iran by Ayatollah Khomeini, pushed production down to 700,000 barrels per day by the end of the decade.[1]

The company had come through all these crises. Investors and other stakeholders – lenders, suppliers, governments – took great confidence in BP's longevity, seeing it as a proxy for enduring excellence and values.

The reason corporations commemorate anniversaries is precisely to instil such confidence among their various stakeholders. The irony was that, in essence, BP was not really an old company at all. BP in 2009 was a company whose culture, structure and leadership had been forged only 20 years earlier, and whose identity was largely owed to one man: John Browne.

The greatest businessman of his generation

Internationally renowned businesspeople, men and women who define their corporations and have reputations extending beyond the sphere of business, are almost always company founders. This is the elite realm of Microsoft's Bill Gates, Apple's Steve Jobs, Oracle's Larry Ellison, Walmart's Sam Walton and even villains like WorldCom's Bernie Ebbers.

The preponderance of technology companies in any list of world-famous businesspeople reflects the fact that software and computer hardware makers are more likely than most to go from being garage-based operations to entering the Fortune 500 in the span of a career. Consumer goods manufacturers and providers of services from investment management to retail also, to a lesser extent, give rise to legendary executives. Again, this is due to high potential growth rates that could allow a founder to go, in a couple of decades, from unlocking the store each morning to ringing the opening bell at the New York Stock Exchange.

Least likely to generate businesspeople of international renown are the traditionally low-growth, asset-heavy industrial enterprises. Manufacturing and natural resources, in which even successful enterprises struggle to achieve double-digit profits growth, rarely give CEOs the opportunity to become household names. Industrial companies are enterprises whose success is the result of applying good management techniques to good assets over a period of decades. Investors want the management of such businesses to deliver steady, low-risk returns, not fireworks. Successful chief executives are usually of the 'evolution not revolution' mould. They're not usually the kind of people who feature on the front cover of *Fortune* magazine or get inundated with requests to appear on CNBC.

Two exceptions come to mind. The first is Jack Welch, who became the very embodiment of a big-business CEO when, after taking charge of General Electric in 1981, he brutally beat a floundering conglomerate into shape, axing workers and product lines, and in the process making GE the United States' biggest and most respected corporation.

The other is John Browne, who enjoyed his superstar-CEO status thanks to his remoulding of BP from a second-tier player in the oil industry – for years seen as a likely takeover target – into the second-largest and most dynamic oil company in the world. His achievements saw him voted the UK's 'most admired' business leader four years in a row – an unprecedented endorsement. It prompted *Fortune* magazine to declare him the most powerful man in business outside America, saw the *Financial Times* dub him the 'Sun King', while the left-leaning *Guardian* declared him 'the nearest thing British business has to a rock star'. His success earned him a knighthood, an appointment to the House of Lords and a fortune worth tens of millions of pounds. In the eyes of the public and most BP employees, he was the very embodiment of BP. Shortly after Browne stepped down as CEO, the company chairman, Peter Sutherland, described him as 'the greatest British businessman of his generation', an accolade that few in the British corporate world would dispute.

Yet what stuck as much in the minds of many of those who worked closely with Browne over the years as his unique skill for grasping complex commercial problems was his emotional detachment, which, on occasion, contributed to fallings-out with long-time friends. Indeed, by the time Sutherland paid him the aforementioned tribute, the two were no longer really speaking. Sutherland had privately voiced his contempt of Browne to mutual friends, while Browne, in his memoirs, refused even to mention Sutherland's name – even when insulting him.[2]

But John Browne had never been an obvious Big Oil CEO. For one, he didn't even drive a car, making him perhaps the only Big Oil executive not to fill up with his own product. Amid an industry of hard men, he was diminutive in stature, with an elfin face and refined tastes. While most peers were of the beer-and-overalls variety, Browne enjoyed tailored suits, opera and the services of a butler. He was also unusual, as a Big Oil CEO, in being gay. His departure from BP followed an unsuccessful legal battle to prevent his young ex-lover, a former male prostitute, from selling a kiss-and-tell story of their time together, a scandal that transformed him from financial news hero to tabloid fodder. Yet it was his sheer detachment from his fellow man – not his sexuality, unrivalled intelligence or sophisticated tastes – that distinguished Browne most for those who knew him best.

All men in high office have an ability to justify, to themselves at least, the inevitable damage they do to others' feelings along the way. Indeed, if one could not bear to inflict the pain of defeat on one's rivals, to punish the incompetent or to dispense with the surplus, one could never succeed in business, or perhaps in any endeavour. Even in this context, Browne was quite exceptional. He was not given to unnecessary unkindness, yet he possessed an apathy towards others that made him, while easy to admire, difficult to develop affection for.

'Remote' and 'aloof' were the kinds of terms newspaper profiles used to describe him.[3] Friends and colleagues I spoke to found him scarcely warmer. Birthdays might be remembered with a card or a telephone call but the recipient was often left with a sense that such acts reflected a well-managed diary, rather than thoughtfulness. If he asked about a subordinate's children, it seemed to some as if it was usually as much about discovering what the younger generation thought about issues such as global warming and the environment. A friend of Browne's who was chief executive of another big British company said of Browne that 'He was the most extreme combination of phenomenally high IQ and minimal EQ,' referring to 'emotional quotient', the rather vague measure in which modern management gurus nowadays put great store.

Despite his exquisite manners, Browne displayed an amazing propensity for human faux pas. When a friend told him he had returned home from a business trip to find his wife had left him, Browne immediately shot back that 'You always have to look forward in life. There's no point looking backwards.' The friend felt less than comforted. 'Two days after your wife has left you unexpectedly, and dramatically,' he said, 'it seemed more than a little trite.'

Two years after his retirement, in an interview in which he discussed the trials of high office, Browne remarked that, in his 40 years at BP, concerns about work had not cost him a single night's sleep. If one was the sort of person to lose sleep over work, he told the journalist, one could not be a CEO.[4]

Over the course of two decades, Browne had slashed tens of thousands of jobs. His cost-cutting was blamed by many for causing oil spills in Alaska – even BP accepted it was a factor[5] – and also the deaths of over 20 workers, and appalling injuries to dozens more, at BP refineries in the United States. But for Browne these were not the kind of things that caused sleepless nights. It was a coldness even the executives at Exxon – traditionally labelled the industry's automatons – could not display.

Twenty years earlier, when Exxon's CEO Lawrence Rawl, widely considered the toughest oil-industry executive of his time, was asked about the criticism he received after the Valdez oil spill, he confessed: 'Sometimes I lay awake at night.'[6]

Petro-prodigy

John Browne was born in Hamburg in 1948. His father was a British army officer of modest background, his mother an interpreter for the occupying allied forces and a half-Jewish Auschwitz survivor. After Browne senior left the army, he took a role with the Anglo-Iranian Oil Company, as BP was then known. This required the Browne family to move to Iran, where the young Browne got his first taste of the oil industry.

As was the practice of expatriate families, Browne was sent back to boarding school in England. His father, in traditional stiff-upper-lip style, held the view that the experience would 'make a man' of his son. Browne's relationship with his father was, as he put it, 'not bad',[7] but his bond with his mother was intensely close and would remain so until she died. Despite the difficulties of separation, Browne thrived at Ely, his housemaster declaring him the brightest pupil he ever taught.[8] The boy's exotic tales of life in Iran made him a novelty to the others, thereby allowing him to avoid the bullying that was prevalent at the school.

Browne went on to study physics at Cambridge, where he developed a reputation for being brilliant but serious: unlike his peers, he never got drunk or misbehaved. Friends assumed this was simply his temperament. While partly true, his sobriety also reflected the fact he didn't want to lose control. He later revealed in his memoirs that he was terrified he might betray the truth about his sexuality if he did. The terror of being found out was partly a reflection of the time. When Browne entered university, homosexual acts between consenting men were still unlawful; though it was decriminalised in 1967, homosexuality remained socially unacceptable for a long time afterwards.

Some friends would speculate that Browne's emotional deafness was related to the fact he bottled up his homosexuality for so long. Others thought it might be due to the influence of his mother's distrust of the human spirit. ('Trust was something my mother, hardly surprising given her background, meted out in small measures,'[9] Browne would later say, referring to his mother's experience of being shunted from ghetto to concentration camp in Nazi Germany.) Others still thought it a feature of his extreme intelligence, which forced him to see the world in unsentimental terms. Perhaps it was all three.

In order to relieve the financial burden of his studies on his father, Browne secured a sponsorship from BP, although he was under no obligation to work for the company after he graduated. Indeed, he was such a brilliant student that his professors assumed he would pursue a career in academia rather than sully his hands with industry. But on receiving his first-class degree, Browne declined the position he was offered on a team studying plate tectonics and joined BP instead, a decision partly clinched by the company's promise to send him to the United States.

As it happened, in 1969, Browne found himself in Alaska. It was not the swinging Sixties American atmosphere that he had envisaged. However, the posting gave him his first experience of petroleum engineering. He worked as part of a team of 20 men who were exploring at Prudhoe Bay on the northern coast of Alaska. BP and two other companies had already made significant oil finds here and the field would turn out to be the biggest ever discovered in North America. Indeed, BP's stake here ensured the company's survival after its expulsion from the Middle East.

In addition to exploration, his time in Alaska also gave Browne his first taste of something else for which he would later become renowned: outsourcing. The team in Prudhoe Bay began generating more data than it could analyse, and Browne realised they needed a computer – still a rarity at that time – to help process it all. But they didn't have the money to buy one. He hit on the novel idea of paying to use someone else's computer in Anchorage, while it was idle, out of office hours. It was his first realisation that a company could access capabilities and technology without actually going to the expense of developing them itself.[10] It was a lesson whose repercussions would resonate for decades to come at BP.

Eighteen months after his arrival, Browne was moved to New York to work on the development plans for Prudhoe Bay. Unknown to himself, he had been marked out as a high-capability candidate and so began a pattern of rapid moves around the company. This programme of constant transition was intended to test him in different situations, to see if he was worthy of – and, if so, to prepare him for – high office.

But the practice had a distinct downside. While Browne's brilliance allowed him to grasp the detail of complex issues quickly, and so to avoid the programme's pitfall of creating jacks of all trades, masters of none, the constant moving robbed him of one great learning experience: time. Specifically, the time to see his plans go wrong and thus be forced to clean up the mess.

By 1980, it was clear that Browne was a candidate for the highest level in BP. He was sent to Stanford University in California to do a Master's degree in management science, which opened his eyes to the latest business theories and practices. But the move also closed the opportunity of a fulfilled personal life to him. Browne's father died a few months before he was due to start the course, and so he suggested to his mother that she come to the US with him. It was a loving gesture, but she was of a generation and background that could not come to terms with the idea of a gay son. While in New York, he had rented a flat in Greenwich Village and frequented gay bars. Now, with his mother present – she lived with him in the US and UK until her death 20 years later – his relationships had to remain clandestine.

Browne tried to broach the subject of his sexuality with her on a couple of occasions but each time she changed the subject. Feeling the pressure from her to follow his peers and settle down, he concocted a story about a love affair with another student at Stanford. He told his mother that he had fallen in love with and had hoped to marry a young woman but that she had ditched him. Alas, while the girl existed, the relationship did not. Nonetheless, until her death, Paula Browne would tell friends the tale of John's lost love, and his heartbroken decision to devote himself to business. When Browne became chief executive, he himself would revive the story for journalists who probed the issue of his bachelorhood.

After spending most of the previous decade wandering the globe, Browne returned to London in 1981. The BP he came home to was a stultifying and highly stratified place. The company's headquarters was a 1960s office block in Moorgate, the second tallest in the City, named Britannic Tower. BP staff called the ugly steel and concrete structure 'West Britannic' to distinguish it from BP's original headquarters, Britannic House, a small Lutyens-designed 1920s building in grey York stone at Finsbury Square.

The Tower had three restaurants, and where one dined depended on seniority. The basement canteen, where staff collected their food on trays, was open to the lower levels of staff. Next door was the senior dining room, known as the 'Coffin Club', with crisp tablecloths and waitress service, which was only open to middle managers. Lunch of a far superior standard was served here, after an aperitif of sherry. The top floor of West Britannic housed 'the chairman's dining room'. Here, the top managers were invited to come each day for a long, leisurely lunch. This started with a drink selected from a long line of bottles on a wooden side table. The chairman at the time, Sir Peter Walters, had once worked in the US, where he had developed a taste for martinis. For him, and others, lunch would begin with a large one. 'Productivity tended to be higher in the morning than the afternoon,' one former executive director who attended the lunches remarked.

The hierarchy extended to tea breaks, which came in the middle of the morning and the middle of the afternoon. Managers of a certain level would be served their tea in china cups on silver trays, prompting the name 'tray men' for senior staff. Many of them had served BP around the globe in previous decades and were working out their time towards retirement. The 'brigadiers', as these over-the-hill tray men were known, were perceived as doing little meaningful work and representing a major barrier to change.

The hierarchical rewards were partly a feature of a UK tax system. Marginal tax rates above 80 per cent made it cheaper to compensate staff in perks than cash. But it also reflected company culture, which still had a colonial feel. 'When I joined in 1980 I felt like I had joined a branch of the civil service,' said David Bamford, who later went on to lead BP's global exploration effort.[11]

BP's other main centre was in Sunbury, south-west of London, where the company's technical brains were based. It was scarcely any more dynamic than West Britannic. Keith Myers started work there after completing his geology degree and felt as though he had not actually left academia. 'The research centre in Sunbury was full of people walking around in lab coats doing work with very little real commercial merit,' he said. 'In the first two years I was there, I felt like I was doing a post-doctorate fellowship. It was like the University of Sunbury.'[12]

Browne was equally disillusioned with the company. Flush with new ideas gleaned from his Stanford sojourn, he knew BP needed change. After losing its Middle Eastern fields, BP was reliant on two areas – the North Sea and Alaska – leading rivals to dismiss it as a 'two-pipeline company'. However, with oil prices around an all-time high in the wake of the Iranian Revolution, there was not the impetus for the kind of dramatic change that Browne thought BP needed.

It would take the catastrophic downturn of the 1980s to create the conditions for such change to take root.

The end of oil?

Oil is famously an industry of peaks and troughs. Oil companies' fortunes and those of their investors, suppliers and employees rise and fall on the price of crude. When prices are high, the streets of Aberdeen, Houston and Kuwait City are crammed with sports cars and expensive German saloons. When oil prices falter, the producing centres of the world see long dole queues and, in some cases, political unrest.

The peaks are exemplified by the four-year period up until July 2008, when US crude prices traded above $147 a barrel – an all-time high, in nominal and inflation-adjusted terms. This period brought riches to those in or connected to the oil business, riches that eclipsed all previous highs. Hundreds of billions of dollars flowed into Saudi Arabia, facilitating the building of glistening new cities in the desert. Soaring oil revenues helped Hugo Chávez mask a defunct economy and export his Bolivarian socialist revolution across South America. It made billionaires of Arab sheikhs and Russian oligarchs, and fortunes of tens of millions of dollars for the CEOs of Western oil majors and the owners of Western oil services companies.

For the biggest Western oil companies it was a time when they literally faced an embarrassment of riches. Exxon, BP and Royal Dutch Shell were already the most valuable corporations in the US, UK and Continental Europe when prices started to take off, but the crude price explosion put even more air between them and their rivals. For years in a row, the companies set records for the highest-ever corporate earnings in their respective jurisdictions. Tens of billions of dollars were paid out to shareholders in dividends. The companies spent as much again buying back their shares because they could not find enough opportunities in which to invest all their cash. The enormous profits prompted accusations that the companies were profiteering at the expense of motorists, and calls for windfall taxes, but clever playing of Western fears about security of energy supply meant this never came to pass.

Ten years earlier, it had all been so different.

Invariably, at every conference on oil, some speaker will refer back to the period from 1997 to 1999, to highlight the oil industry's complementary propensity for downturns. Over this period, crude dropped from over $26 a barrel – well above its long-term, inflation-adjusted average of around $18 a barrel – to under $10 a barrel. On an inflation-adjusted basis, prices hit their lowest level since before the first oil shock in 1973, when an emerging Organization of the Petroleum Exporting Countries (OPEC) had flexed its muscle and hiked the price of its crude by 70 per cent overnight. I was working as an oil broker at the time and regularly heard clients predict prices would fall to $1.50 – just above the marginal price of extracting crude from Saudi Arabia's deserts. The late 1990s were, by any standards, a nightmare for the oil industry.

But when you ask lifelong oil men about the industry's lowest point, they go back another decade. In the early 1980s, oil prices collapsed after a decade of price peaks that had prompted a tsunami of investment into new fields and refineries. A five-year trough in prices from 1982 onwards led to a period of bloodletting that is seared in oil veterans' minds. 'I remember the crash in oil prices in 1982–83 when we lost 400,000 people out of the industry in the United States. It was painful,' Exxon Mobil Chief Executive Rex Tillerson told me in an interview in 2008 when I asked him why, with oil prices around $147 a barrel, his company was still investing cautiously. 'I had to let a lot of those people go. And those things, you don't lose sight of them easily,' he added, with a hint of emotion uncharacteristic of Exxon's public comments.

Exxon weathered the 1980s downturn better than most. It was the largest and most self-assured of the Seven Sisters, as the biggest Western oil companies, who until the 1970s had controlled global oil production, were known.

Across the Atlantic, the suffering was equally felt by Royal Dutch Shell, the second largest of the Sisters. Lacking Exxon's absolute certainty in its way of doing things, the downturn prompted an existential crisis at the Anglo-Dutch oil company.

Shell had for decades been organised according to what it called a 'matrix' structure, which basically meant that the company was broken down into regional units. Some of these, especially the US unit, the Shell Oil Company, enjoyed a large amount of autonomy. But all the units were also managed along functional lines, which were operated out of group headquarters. This meant that every manager had two bosses: the head of Shell in the country in which he or she worked, and the global head of their function, be it exploration, refining or fuel marketing. So a driller in the North Sea answered to the head of Shell UK but also to the global head of exploration. A refinery manager in Le Havre reported to the head of Shell France, and also to the global head of refining. It was a complicated system, made even more so by the tangled corporate structure of the group. What most people considered 'Shell' was actually a virtual concept.

The group was 60 per cent owned by a Netherlands-listed holding company, Royal Dutch, and 40 per cent by a British one, The 'Shell' Transport and Trading Company. Each of these companies had its own board and shareholders and they regularly wrangled with each other for control of the group. To keep both sides happy, the Shell group operated out of two headquarters, one in London (where the head of refining, for example, sat) and one in The Hague (where the head of exploration sat). Consensus was the name of the game at Shell. There had to be consensus between Shell T&T and Royal Dutch to make any corporate decisions, and between the country and function heads to make any operational decisions. It was bureaucratic. It was slow-moving. It was management by committee.

By contrast, Exxon had a strict, centralised and rules-based management system. Directly or indirectly, everyone reported into head office in Irving, on the outskirts of Dallas, Texas. Function, rather than geography, was the primary organising principle. Indeed, national units' existence was often little more than a legal technicality, undertaken to fulfil local tax and other regulatory obligations. Irving set procedures that dictated every decision made by every Exxon employee in every corner of the planet. Compliance was not open to interpretation or negotiation. Rivals said the structure made Exxon risk-averse and derided its staff as androids who were incapable of thinking for themselves, even when it came to dressing. Exxon employees invariably turned up to meetings in dark blue suits and red ties, so the joke went. But the absolute clarity Exxon employees enjoyed with respect to what was required of them ensured the best efficiency in the industry. 'If you wanted to open a bank account in Tokyo, you had to ask Dallas,' the chief executive of one big rival once quipped to me.

Shell had in the past mulled over changing its structure to improve performance. It had considered the Exxon model but always decided this could not be replicated, at least not by a company with as convoluted an ownership structure as Shell. The question was: should they go in the opposite direction? In the early 1970s, and again in the late 1970s, Shell invited management consultant McKinsey in to conduct studies on the group. McKinsey said that, while Shell's structure, by encouraging consensus, had fostered technical prowess, it was not best suited to maximising profits. They suggested that by giving more independence to operating units – essentially allowing them to operate as stand-alone businesses – Shell could increase entrepreneurship and, therefore, profits. Individual assets, such as refineries and oil fields, could be allowed to set their own strategies, arrange their own procurement and decide on their own engineering solutions. Managers of the independent units would be released of the responsibility to reach consensus with the various other internal stakeholders before making decisions. Instead, they could simply be held accountable to meeting profit targets.

While ostensibly a creative solution, McKinsey was only doing what management consultants usually do, that is to say, sell ideas gleaned from previous clients to new ones. In this case McKinsey was pushing an idea that its client General Electric had pioneered in the late 1960s and 1970s. Indeed, the idea of organising one's corporation into strategic business units had, by the late 1970s, become a fixture in teaching manuals at business schools across America. Nonetheless, it was a revolutionary proposal for an oil company, where a high and consistent level of technical expertise was required across the corporation, to ensure its volatile product did not realise its potential for disaster.

In the event, as unoriginal as the idea was, it proved too daring for the change-resistant Shell. The McKinsey reports, after being dutifully discussed, were left to gather dust. In 1986, however, with oil prices at a 13-year low, the management consultants were invited in again.

This was the 1980s, a time when Michael Douglas was declaring that 'greed is good' in the movie *Wall Street* and when Margaret Thatcher was rolling back state involvement in the British economy, setting a model that Europe would later follow. 'Shareholder value' was becoming the dominant management philosophy. Companies might have responsibilities to their employees, customers and the communities in which they operated, but these should be met only in so far as doing so facilitated the corporation's primary role, which was maximising shareholder returns.

Predictably, McKinsey again suggested ditching the 'matrix' structure – which the consultancy had itself devised for Shell in the 1950s – in favour of the decentralised business model.

Shell's boss, Lodewijk van Wachem, a strict Calvinist, was under no illusion about the difficulties they faced in a world with oil prices below $10 a barrel. However, van Wachem personified Shell's conservative and cautious culture. He doubted the organisation could handle the kind of cultural change that the new plan envisaged. Even if profitability was boosted, would it be sustainable? Did Shell risk losing the technical excellence that made it the best oil explorer, and one of the best asset operators, of all the big oil companies? The more van Wachem and Shell's directors mulled it over, the more they felt unable to bring themselves to adopt the glitzy business model advocated by the management consultants. Yet again, the consultants' report was left to gather dust.[13]

A couple of miles north-east of Shell's London HQ was Europe's second-largest oil company, BP, also organised under a matrix structure. If Shell was finding the 1980s downturn a challenge, BP was finding it an existential threat. Chairman Peter Walters had tried to reduce the reliance on the North Sea and Alaska, which were expected to run dry in the 1990s, by building a diversified portfolio of assets. But his bias towards political stability, rather than prospectivity, was a critical flaw. BP explorationists like David Bamford could hardly believe it when they heard Walters talk about sending geologists to explore in France, Germany, Spain and Switzerland. 'It was complete geological nonsense,' he said.[14]

Yet for Walters, after the searing experience of BP losing its core Middle Eastern assets in the 1970s, conservatism was the name of the game. 'It's better to invest in a mature business we know a lot about than be seduced into fast-growing businesses where our chance of emerging as number one or number two is problematical,' he once told an interviewer.[15]

There were some positive changes under Walters. He sent Bob Horton, who had distinguished himself as an ardent cost-cutter in BP's chemical division, over to Cleveland to turn around Standard Oil of Ohio, or Sohio, the US oil company in which BP had picked up a majority stake in the early 1970s. Horton took along the rising star John Browne to help him, and their efforts were so successful that BP decided to buy out the minority shareholders in 1987.

The same year, the British government gave up its controlling stake in British Petroleum in a share flotation that was supposed to be the crowning glory of Margaret Thatcher's privatisation drive. Royal

Marines abseiled down the side of Britannic Tower to unveil a banner showing the price at which shares would be offered, in an attempt to drum up public interest. This should have been the start of a bright new future for BP, unshackling it from the last vestiges of political interference. Yet the privatisation left the company with a crippling burden.

In the days before the planned share sale, the stock market collapsed on what became known internationally as 'Black Monday'. Investors who bought the shares soon found themselves nursing losses. It was a major embarrassment for Thatcher, and a setback in her aim to create a culture of share ownership in Britain.

Others saw the drop as an opportunity. The Kuwaiti government's investment vehicle swooped in and snapped up over 20 per cent of BP. Senior executives worried the Kuwaitis would try to buy control. Replacing a democratic, pro-capitalist government with an authoritarian monarchy as its majority shareholder would hardly represent progress for the company.

In the event, they need not have worried. Margaret Thatcher was having none of it. At the risk of selling a few fewer jet fighters to the Arab kingdom, she forced the Kuwaitis to reduce their stake to 10 per cent.

The bad news for BP was that it was compelled to buy the shares. Lacking the money to do so, it had to turn to its banks, to whom it was already heavily in hock after spending $7.5 billion to buy outright control of Sohio. The result was a heavily indebted BP wading its way through a period of sluggish oil prices: not a recipe for growth. Repeatedly in the coming years the company had to reassure investors that it had enough cash to meet its financial obligations. Finance Director David Simon continually batted back suggestions that BP would have to go to shareholders and ask for money to bolster its balance sheet.

However, even if BP didn't go bankrupt, the question remained as to whether it could offer investors any prospect of attractive returns. With so little cash to invest, it would be hard to expand production. If a corporate raider came along, would investors stick by management? Top managers including Simon and others were not sure they would. And while Thatcher wouldn't allow a Middle Eastern government to buy BP, there was no reason to believe she would block a takeover by a US or European rival. Indeed, rumours circulated that Shell or even the Norwegian state-controlled Statoil, a relative newcomer, would mount a bid.

By the end of the 1980s, it was clear to all at BP that serious change was needed. The question was: who would lead it? Horton's success in turning around Sohio solidified his reputation as the natural successor to Walters. This left the question of his team. Following the Sohio takeover and the merging of Sohio's oil fields with BP's US assets, Browne was made head of exploration and production (E&P) – also known as the 'upstream' division – for North America. As Horton moved up to the top job, the role of head of E&P for all of BP became vacant. Even with low oil prices, this unit, which found and pumped the oil and gas, was BP's main profit driver. But it was hopelessly inefficient compared to rivals. It was clear that the company's recovery depended upon turning E&P around.

On paper, Browne was the clear choice for the job. But his superiors worried about what had never been written down.

Not sexist enough

John Browne's proposed appointment as head of E&P would elevate him to the main board of the company. British Petroleum, as it was still officially known, was an august British institution. It was the UK's largest company and its every move was closely watched by the media, politicians and the broader public. Incompetence could be tolerated. Scandal was an altogether more serious affair.

Browne was convinced no one knew about his sexuality. He had always been discreet about his romantic involvements. Of course, living with his mother and working up to 18 hours a day, seven days a week, limited the opportunities for a hedonistic existence, although the fact that Paula Browne acted as her son's escort to corporate events did cause some mutterings. After all, being 'close to his mother' was an old euphemism for being gay. Yet, in so far as they thought about it, most people at BP did not draw this conclusion. Colleagues such as Tom Hamilton, to whom Paula had told the story of Browne's lost love, accepted that he had simply neglected that part of his life. Mindful of his unemotional ways, others thought that he might simply be 'asexual'.

Those who had risen through the ranks with him, however, had divined the truth. It wasn't anything specific, rather a combination of small things: not sharing sexist banter with colleagues; his general disinterest in women. 'When six of you are sitting in a room and a pretty girl walks in and only five turn around, you notice,' said one close colleague from the period.

The board decided that Browne was their man for exploration and production, but before the appointment was confirmed Peter Walters approached one of the younger managing directors who knew Browne well and asked if his sexuality would be a problem. Only after Walters was assured that Browne was discreet and unlikely to cause scandal was the young executive's position as head of E&P confirmed.

The appointment at just 40 years of age, combined with his mentor Horton's elevation to the chief executive and chairman role, provided the opportunity for Browne to execute his plans for BP. Some at BP argued the company could survive by being 'nimble number three'. It was half the size of rivals Shell and Exxon and some thought it could take advantage of this fact to be faster moving.

But Browne's ambition was not to be sated by coming third.

Ditching the matrix

Browne had long detested BP's 'clunky' matrix structure. He blamed the various checks and balances it involved for slow decision-making and a lack of entrepreneurship.[16] He knew about the ideas McKinsey had pitched to Shell, and, unlike the Anglo-Dutch oil giant, he wasn't concerned about whether BP's culture could handle the upheaval. Shortly after he was appointed head of E&P, Browne moved his office out of Britannic Tower into a much smaller building a short distance away at Blackfriars. He was going to start a revolution and he didn't want any of the brigadiers of West Britannic gazing over his shoulder.

Browne appointed Rodney Chase, an imposing executive who had worked with him when Browne was group treasurer, as his deputy. Tom Hamilton, a former Exxon geologist who had worked with him at Sohio, was made head of frontier and international exploration. Rodney would become the enforcer who helped knock the existing portfolio into shape, while Hamilton scoured the globe for new opportunities.

Browne decided BPX, as the E&P division became known, would be unlike any previous upstream division. Indeed, it would, in essence, stop being an oil company altogether.

Under BP's existing matrix structure, a manager of a big oil field would have to work through functional departments and geographic leadership to execute his project. He would have to agree his exploration plan with the drilling department, work with the project development department to come up with a design for whatever offshore platform or structure was being built and co-ordinate with the country or regional manager to ensure that his plans fit with other BP activities in the region.

Under Browne's plan, big oil fields or clusters of smaller fields would become independent companies. The boss of each would function like a chief executive, unobligated to go through BP's central bureaucracy to get things done. Indeed, the central bureaucracy would largely be dispensed with, as the project manager was encouraged to look outside the company for technical expertise such as seismic surveys or the drilling of wells.

The thinking was that outsourcing would lead to greater efficiency. BP's drilling department had little incentive, the logic went, to be efficient because it was essentially supplying a service to a captive customer: BP. By buying in services, one let the free market instil efficiency and push down costs.

The field manager would make all the decisions on the design of the project, the purchasing of the items needed and ongoing operations. Field managers would even manage their own public relations. Very little would be done centrally. Rather than manage oil fields around the world, Browne and his department would simply act as a fund manager. BPX's primary role would be to allocate capital, deciding which projects to back, but then stepping aside and letting the project bosses run their own businesses.

The only management that Browne, at the centre, would provide was to draw up performance contracts with the business unit leaders, or BULs (pronounced 'bulls', a title that fit with the macho atmosphere that soon developed), and hold them to the contracts.

As a highly analytical thinker, Browne liked numerical targets – but he thought having too many of them confused people. In the end, the performance contracts centred on two principal targets: production levels and cost. Since profit was largely determined by the oil price, which no one could influence, it seemed pointless to measure units on profitability.

Browne's plan appealed to Chase's aggressive nature but Hamilton, the affable Ohioan, saw an obvious flaw. If managers had incentives to hit a limited number of short-term targets, the chances were they would focus on these, potentially to the exclusion of equally important activities that were not being measured, such as maintaining equipment in good, safe condition. 'I told him you've got to be careful,' Hamilton later remembered. 'If your barn needs to be painted and you look at it and you say, "Well, it probably needs to be painted, but it can wait until next year," that's what's going to happen. People are going to push off costs that show up on the bottom line.'[17]

Browne was not convinced. One of the big problems at BP, as he saw it, was that the engineers who called all the shots wanted to 'gold plate' every platform, refinery and pipeline. Nonetheless, as a measure to tackle such concerns, Browne said he would appoint regional vice presidents with responsibility for ensuring managers adhered to BP's environmental, safety and ethical standards.

Having settled on the structure, Browne had to decide what kind of portfolio it would manage. Walters' strategy of diversifying into an array of safe but oil-poor countries had been a failure. It left BP with a basket of non-material interests: a stake in a small field here, a stake in a small field there, often in areas with little growth opportunity. Only two material pieces stuck out – Alaska and the North Sea – and both were in decline. In the future, Browne decreed, BP would focus on big fields in areas that offered the prospect of material future discoveries.

A year after Browne was appointed head of BPX, his plan was ready to be unveiled. In March 1990, he invited his top 100 executives to a week-long conference. It would be held at the Phoenician, a luxury resort just outside Phoenix, Arizona. On the face of it, the Phoenician was a strange place to host a conference on how to tackle BP's high cost base. The resort had recently been completed at a cost of almost $300 million and was set in 130 idyllic acres at the base of Camelback Mountain. It had a golf course, nine swimming pools – tiled with mother-of-pearl – and several restaurants, serving delicacies such as fried rattlesnake with Cajun jalapeño crème fraiche. But all was not as it seemed. Shortly after the Phoenician opened, the owner went bankrupt, and the Savings and Loan bank, which financed it, had to be bailed out by the government.[18] Now, to keep the hotel afloat, rooms were being heavily discounted. There was one afternoon of golf, and some of the men skipped into Phoenix at night to check out the strip bars, but if anyone thought they were in for a luxurious corporate junket, they were sorely mistaken. The attendees were locked in conference rooms for long days of sessions run by management consultants. They had to determine how to put Browne's new plan into practice.

Browne made it clear that fields that didn't fit his strategy would be sold, so everyone knew they had to fight to keep their projects, and themselves, in the company. One by one, people stood up and outlined how profitable their field would be. Then they were challenged. As some were torn to shreds by their colleagues, tears formed in heavy men's eyes. 'It was mortal combat in terms of how we were going to allocate funds,' Hamilton later recalled.[19]

In addition to deciding which fields to jettison, the meeting also sought to define the extent to which decentralisation should be pushed. The proposals ranged from what was termed the 'Alaskan view', which envisaged a central BPX function with as few as 25 people, to the 'Sunbury view', named after BP's old-school south-west London facility, which envisaged a body of over 1,000 people still offering technical guidance to asset managers.

By Wednesday morning, the debate was leaning in favour of the Alaskan view. It was all too much for the chief driller, Cliff Simpson. He had spent his career honing his specialism and saw himself as the guardian of good drilling practice at BP – the man who ensured BP drilled good, safe wells. He stormed out of the meeting. Two days later, after the debate had settled on the minimal overhead model, he returned and castigated those present. 'I understand the mood of the room but basically, you have destroyed a lifetime's work of many people,' he declared, according to another person present.

The message from Phoenix was unambiguous: BP was no longer a technical company but a commercial one. Its primary role was to decide which projects to back, and where possible have others do the actual work.

The writing was on the wall for people like Simpson. At Exxon and Shell, the role of chief driller was one of the company's most powerful and specialised. He or she was an experienced driller who managed the company's drilling operations through a direct chain of command. Under Browne's new regime, the position of chief driller would continue at BP but it would be filled by commercial managers on short rotations, whose main remit was to help field managers drive down the cost of hiring outside drilling contractors. It wasn't until after the United States' worst ever oil spill 20 years later that BP once again appointed an experienced driller to manage its global drilling operations.

Browne's changes blew like a whirlwind through BPX. BP had become bloated and inefficient by 1990. Horton told staff it needed to do 'more for less' and Browne's new structure more than met the challenge. BP halved its workforce from almost 100,000 to just over 50,000 from 1992 to 1996. The old values like loyalty and paternalism were jettisoned and a harsher atmosphere descended on the company.

Yet for many of the employees who stayed behind, the changes were exhilarating. BPX was broken into around 50 different strategic business units. Managers in the units felt unshackled from bureaucracy and in control of their own destinies. The new atmosphere fostered entrepreneurship and people suddenly began proposing ideas that went against the traditional way of doing things.

An example was the Andrew Field in the North Sea, which had been discovered in 1974 but lay dormant for almost 20 years because BP was unable to find a way to develop it economically. In the early 1990s, the field's prospects were revisited by Colin Maclean, a geologist-turned-commercial manager who had left BP in the early 1980s because of its stifling culture, only to find himself back at the corporation after it bought his next employer, Britoil, in 1988. Maclean looked at the Andrew Field and hit on an idea: BP should offer a new kind of contract to the companies that built the equipment needed to exploit the field. In the past, such companies quoted a fixed fee to deliver a platform. Platforms were so enormous and so expensive that the construction companies who built them could not afford to bear the risk of cost overruns. Consequently, they had to build a contingency into the price. What Maclean suggested was that BP offer a contract that excluded such a contingency. Instead, if the contractor ran over budget, BP would cover around half the overrun. Similarly, if the contractor could deliver the platform under budget, BP and the contractors would share the savings.

Browne jumped at the deal and so did the contractors. The business model was soon adopted across the North Sea, which at the time was struggling to attract investment. 'That was the reinvention of the North Sea,' Maclean said. 'It was a visible outcome of the way John had us think.'[20] The Andrew Field also displayed Browne's ability to spot a political as well as a commercial opportunity. When it came to inaugurating the field, it was the leader of the opposition, the Right Honourable Tony Blair, whom Browne invited to cut the ribbon, not a member of the Tory government. The Conservatives were enraged but Browne didn't care. New Labour was on the rise and Blair would soon be in Downing Street.

Turtles

Colin Maclean had started out as a technical expert but transformed himself into a professional manager. Over the coming decade and a half, he would trot around the globe at Browne's behest, running refineries, managing procurement and helping orchestrate mergers. At one point he owned five houses in four countries.

It was clear that to get ahead in BP one would have to reinvent oneself as a commercially focused person. For ambitious young engineers like Andy Inglis, who would later himself become head of BP's exploration and production unit, it meant reconsidering one's career trajectory. 'I joined BP in 1980, and in 1990 was told that Mechanical Engineering was not considered core to BP's strategy, that we would follow a track of outsourcing and use of the contracting industry,' he said. 'This caused me to broaden into other disciplines and areas.'[21]

For those who were successful, the opportunities kept coming. Whereas Exxon and Shell believed in keeping people in roles for years, gradually promoting them up a functional hierarchy, BP favoured rapid moves. The logic was that a good business brain could manage a refinery as well as an oil field or a chemical plant. And since there wasn't perceived to be any benefit in gaining specialist knowledge, the successful could move on to greater things every other year.

For the most ambitious of young engineers, Browne opened up a path to future greatness. When he became head of BPX, he created a new position of executive assistant. At first, people didn't know what the role involved, and suspected it would simply mean acting as a glorified secretary, carrying Browne's briefcase around the world. Indeed, the position did have its menial tasks, including deciphering Browne's illegible handwriting for managers and making sure he was well stocked in his favourite wine, the white Burgundy Puligny Montrachet, and in the El Rey del Mundo Cuban cigars he enjoyed. But it also provided the best business training any of its holders ever experienced. The executive assistants effectively became Browne's right hand, following him around the world, sitting in on every business meeting, passing on orders to division heads and even checking that managers had indeed followed those orders.

The executive assistants came to be known as 'turtles'. When reporters later heard of the nickname, it was ascribed to the cartoon strip *Teenage Mutant Ninja Turtles*, about a group of genetically enhanced turtles tutored in martial arts by their ninja master. The actual reason for the tag was more literal: Browne covered hundreds of thousands of miles each year, so following him around required the young executive assistants to 'travel with their home on their backs'.

The first turtle, even before the nickname was applied, was John Manzoni. From an affluent Birmingham family, Manzoni's aggressive, hard-driving temperament was honed on the rugby pitches of his boarding school and later Imperial College London. He graduated with a degree in civil engineering and joined BP in 1983. For the next seven years he worked as an engineer on North Sea projects, establishing a reputation as a workaholic with fierce ambition. Five a.m. visits to the gym helped maintain an imposing physical presence (one former colleague admiringly described him as 'built like a brick shithouse'), which contributed to the sense of menace he often projected. While some challenged the common accusation that he could be a bully, even his supporters found it hard to depict him in an amiable light. 'You couldn't classify him as a nice guy,' remarked one business unit leader, 'but you could admire him for his cold steel.'

Despite Manzoni's impressive intellect and capacity for hard work, Browne soon realised that one executive assistant could not keep up with his enormous workload. Less than a year later, he appointed a second turtle.

Tony Hayward, one of the junior managers invited to the Phoenix meeting to help organise the proceedings, had impressed Browne sufficiently to be invited to become the second turtle. As one of seven children from parents of modest means, Hayward had been state-school-educated, unlike many top BP managers. He came from Slough, the commercial conurbation west of London whose charms can be discerned by the fact that Ricky Gervais chose it as the setting for TV comedy *The Office*. Aged 33 and head of a small team of geologists, Hayward was unsure whether to accept what was still an unknown role. His line manager advised against it, suggesting that if he hung in where he was, he might in a few years be promoted to the line manager's role, and if he really played his cards right, he could even end up as chief geologist some day.

The line manager had underestimated Hayward's ambition, however, and his ability to sense which way the wind was blowing. Hayward accepted Browne's offer and later joked about his line manager's advice, noting that the chief geologist role was one that BP in time abolished.[22]

Hayward and Manzoni had little in common. Hayward was shy and often uncomfortable in social groups, whereas Manzoni was more polished and self-confident. Manzoni, like Browne, had an analytical approach to business. He liked to trawl through reams of facts and figures to arrive at a conclusion. While decisive, he was ready to change his mind if the facts changed. Hayward was a more intuitive business thinker and reluctant to change his mind once he had made a decision. Fortunately for him, his intuition was usually right.

Proximity would breed friendship between the two young executives, but their twinning as turtles also marked the beginning of a decade and a half of intense rivalry. It was clear that Browne's next role would be chief executive and, in both men's minds, the race to succeed him in that job had already begun.

Into the limelight

Browne didn't have to wait too long for his own destiny to be fulfilled. Bob Horton had started the cost-cutting and increased focus on performance that BP was perceived to need, but he had also bet on the oil price remaining high, tailoring investments accordingly. In the event the oil price fell, forcing the company to cut the dividend in 1992. Horton might have survived this but, while the board had backed many of his strategies, it found it could not back him now; over the course of two years, he had managed to alienate most of the top executives with what many saw as his abrasive and arrogant management style. In an immediately infamous interview he gave to *Fortune* magazine, he noted that, 'Because I am blessed by my good brain, I tend to get the right answer rather quicker and more often than most people.' Unsurprisingly, the board saw this as yet another sign of Horton's incorrigible arrogance. They decided he had been corrupted by his time at Sohio, developing the presidential affectations the British associated with and despised in American CEOs, and ousted him.[23]

Horton was succeeded as chief executive by the gregarious David Simon. In line with what was increasingly being deemed in the UK to be best corporate governance practice, the chairmanship was spun off and made into a non-executive role. Board member John Baring, of the banking dynasty and better known as Lord Ashburton, was made non-executive chairman. Ashburton had no day-to-day operational authority. He and the other non-executives were there to ensure that the management adhered to the business's broader strategic goals and always acted in shareholders' interests. The non-executives were not supposed to second-guess the executive management on operational or lesser strategic issues.

David Simon had come from the supply and trading side of the business and so was happy to leave Browne to his own devices with BPX, even though it constituted most of the company. Two years later, in 1994, Simon walked into Browne's office and announced that the board had decided Browne should succeed him in 1995 and that Simon would step up to replace Ashburton. Rolf Stomberg, a board member at the time, later said, 'There was no second contender. That was something new. Between David and Bob it was a two-horse race. But not with John.'[24]

The appointment thrust Browne into the limelight and his unusual personal circumstances soon became the subject of much speculation. One newspaper wrote that 'he may be the head of Britain's biggest oil company but Sir John Browne is still a bit of a mummy's boy at heart', while another did a profile under the headline: 'BP, mother and me'. Several interviewers asked him outright if he was gay. 'You have got the wrong man there,' he told the *Financial Times*, who published the comment. Others, including *The Times*, declined to print Browne's denials, deciding that to do so would be unfair, as it would only give credibility to the rumours.

Should Browne have been honest and come out about his sexuality? Friends and colleagues from the time said it would have made his job impossible. He was in the middle of leading one of the most brutal turnarounds in British corporate history in a macho industry. And it was about to get a lot more brutal.

Internally, the promotion gave Browne the opportunity to roll out his beloved 'hub and spoke' system of strategic business units across the company. Just as big oil fields were spun off into independent enterprises, refineries were established as business units. The restructuring of BPX had led to a massive improvement in financial performance and the expansion of the process to the whole company, and associated cost-cutting, continued to drive strong results through 1995 and 1996. Analysts and investors were suitably wowed.

'BP's common stock has been the best performer in the international integrated oil group the last two years,' raved analysts at investment bank Morgan Stanley, 'as the growth rate of earnings has surpassed every company in the peer group. The drivers: a major rationalization of its business mix, rehabilitation of its financial position, more optimal allocation of capital, and a secular revamping of its corporate culture.'

Management experts were also blown away. Scholars from Harvard, Stanford and other business schools came to study BP's new business model and wrote glowing verdicts. Browne's use of peer groups – forums for managers with similar types of assets to share tips and experiences – was seen as especially mould-breaking. They weren't an altogether new idea but BP was commended for being uniquely successful in their employment – and certainly Browne had had to make some concessions to information-sharing in his new centralised business structure. The inherent shortcoming of an organisation with a skeletal central management is that it makes it harder for lessons learnt in one part of the business to be shared elsewhere. 'BP's peer groups were a powerful and effective means of achieving mutual co-operation and learning,' one management expert wrote in the *Long Range Planning Journal*, one of the leading international journals in the field of strategic management. The *Harvard Business Review* said the peer group system ensured that managers overcame the inclination to focus on their own narrow goals and also co-operated to achieve broader company objectives: 'Unit managers are equally committed to both their company's overall success and their unit's performance.'

BP became seen as the epitome of the 'learning enterprise',[25] whose very structure ensured that it constantly improved, constantly learnt from mistakes and adapted to the outside environment. The heads of Fortune 500 companies voted BP second only to Microsoft as the 'most admired knowledge enterprise' in the world, in a competition previously dominated by tech companies.[26]

Browne lapped up the praise from the experts – people he knew really appreciated what he was trying to do. They understood that he was not simply a great business manager but truly a business creator. 'Our strategy is our organisation,' he declared. 'In the old days companies had the luxury of stable, long-lived strategies. No longer; strategies continue to shift to reflect changing market conditions. Organisations that succeed have structures that are as supple and adaptable as the strategies they reflect.'[27]

In time, everything the so-called management experts praised – the peer groups, the independent business units, the absence of internal audit functions – would be ditched. They simply didn't work.

The dealmaker

Shortly after Browne was made head of BP's upstream division in 1989, he had delivered a speech at a gathering in Tokyo. 'The number of truly global players seems certain to decline,' he said. 'I would not be surprised if in ten years there were three sisters, instead of seven.'[28]

At the time, BP was seen as a potential takeover target because it was in such a spectacular mess but analysts did not predict broader industry consolidation. Indeed, the sheer scale of the companies involved was seen as precluding this. Browne knew better. The question was: would BP be a consolidator or prey?

Not long after he was made chief executive, Browne received a call from his opposite number at Shell, inviting him for a chat. Browne read the invitation as code for entering talks on a tie-up. Given Shell's size, it was clear that BP would be the junior partner. Browne was not about to throw in the towel just as he was made boss. He strung Shell along for a while with no conclusion, which was the conclusion he wanted.[29]

By 1996, BP had recovered sufficiently to think of itself as an initiator of deals rather than a target. Browne set his eyes on Mobil, the second-largest US oil major, whose portfolio of assets he thought was the best fit for BP.

He was rebuffed. Unfazed, he immediately turned to the second company on his list: Amoco, the fourth-largest US oil producer. In Amoco, Browne found a more willing target. In 1992, Amoco CEO and chairman Larry Fuller had called David Simon on his first day as chief executive of BP, proposing a merger in a deal that would have given Amoco 70 per cent of the enlarged group. Simon had rejected the advance. In the ensuing years, Fuller had remained convinced of the need for a deal but his bargaining position had weakened: he approached Mobil but was turned down; talks with Chevron revealed that the California-based company would insist on getting rid of Amoco's Chicago headquarters and provide minimal board representation for Amoco directors. When Browne appeared, offering a higher price, promising to keep the Amoco skyscraper as a joint HQ and proposing parity on the board of the enlarged group for BP and Amoco representatives, Fuller went for it.

At the last moment, BP reneged on its offer of joint board representation but Amoco still agreed. The oil price kept falling and Larry Fuller didn't see that he had many options.[30]

In August 1998, the two companies announced what, at the time, was the largest industrial merger ever, to create a company with a market capitalisation of $110 billion. While billed as a merger of equals, BP shareholders would own 60 per cent of the enlarged group and the primary listing of the group would be in London. Browne would be CEO while Fuller and Peter Sutherland, by now non-executive chairman of BP, would become joint chairmen. Most of the top jobs went to BP executives.

BP's leadership barely disguised their contempt for their new 'partners'. BP managers saw themselves as international people and their partners as hailing from the boondocks, or as one BP director said, '"Bubba" land. It's Ku Klux Klan territory. When you looked at the Amoco people who were appointed to the board, half of them had not been abroad before. They didn't even have passports.'

Across BP, the Chicago-based company was derided as Midwestern, parochial, and, worst of all, lacking in entrepreneurship. Amoco had been a process-driven, hierarchical company. Its employees were expected to follow orders, unlike their commercially savvy opposite numbers at BP, who were allowed to make up their own rules, provided they delivered the desired results at the end of the day.

A year after the deal was signed, BP broke its word and shuttered Amoco's Chicago HQ. By the following year, few of the Amoco senior managers remained. Fuller left embittered and embarrassed at the dismemberment of what had been a Chicago colossus. He left the city.

BP's executives did not give much thought to what was lost as Amoco's processes and people walked out the door. The Amoco culture had fostered deep technical expertise – something overlooked for almost a decade at BP, and which the company was increasingly in need of. Such cultural strengths pay off in the long term, however, and Browne had his eye on short-term goals. The Amoco takeover was not premised on accessing technology, leveraging brands or any of the other growth-based reasons companies have for making acquisitions. It was simply about cost. Oil was at $10 a barrel and only those who were big and who could achieve economies of scale were seen as capable of surviving.

'Synergies', or cost savings, would add $2 billion a year onto earnings within two years, Browne declared. If this all sounded rather short-termist, the CEO was not apologising for it. 'I am a great believer in short-term profits,' he said. 'A period of 20 years – the normal horizon on which an oil company plans – is just 80 quarters.'[31] He ordered his managing directors, the heads of the various divisions, to take out the knife. 'Go to the limit,' he told them. 'If we go too far we can always pull back later.'

Thousands of jobs were cut, many of them via offers of early retirement, to the extent that, by the end of the 1990s, the company's average retirement age had fallen to 52. Managers' remuneration became increasingly tied to achieving the targets in their performance contracts. Bonuses, if earned, were supposed to account for 15–20 per cent of junior supervisors' pay and up to 70 per cent of top managers' basic pay.

The brutality of the integration process and the performance-related remuneration packages meant only ambitious, ends-focused people felt at home at BP. Managers who cared about means and broader measures of performance beyond production and cost levels increasingly did not see a role for themselves at the company. Browne and his division heads did not appear to question whether the company was losing valuable wisdom, or whether the average BP worker was becoming someone with a higher risk tolerance than might be healthy in a hazardous industry.

Browne's move on Amoco had shaken the industry. The outlook for oil prices was dismal, with Kuwait predicting $5 a barrel oil prices and *The Economist* running a front cover declaring the world 'Drowning in Oil'. (Years later, after oil prices soared towards $100 a barrel, speakers at oil conferences would regularly use blow-ups of *The Economist* cover just to get a laugh.) BP's rivals immediately saw that if they wanted to survive in this new industry landscape, they would have to get bigger quickly. Exxon made a bid for Mobil, and was successful. France's two oil majors, Total and Elf, merged, also incorporating Belgium's Fina. Chevron merged with Texaco and Conoco bought Phillips. Only Shell remained above the fray. With oil prices on the floor, companies had no cash to buy rivals, and instead offered targets their own shares. Shell, with its complex ownership structure, found it impossible to issue enough new shares to buy a rival of any size.

A few months after sealing the Amoco deal, Browne followed up by buying another big US oil company, Arco, for $27 billion. The deal saw BP leapfrog Shell to become the world's second-largest oil company behind Exxon Mobil, as the industry leader was now known. As it had done with its own portfolio, BP broke Amoco and Arco's assets into strategic business units.

With this substantially enlarged portfolio, it was considered unwieldy to continue with the existing system whereby regional presidents looked over the business unit leaders' shoulders to ensure they were following BP's environmental and safety policies. Browne decided to push responsibility for ensuring primary compliance with these policies down to the BULs themselves. From now on, oil field and refinery managers would regulate themselves in relation to health, safety and ethical matters.

There was one major problem with this: it is hard to measure good environmental and safety practices. This is largely because a failure to comply with best practices is not immediately evident. One ought, for example, to prevent oil spills from pipelines by preventing corrosion – but the very gradual appearance of rust means any one manager, who knows he isn't going to be in that particular role for more than a year or two, can save costs by putting off maintenance until the next guy comes along. Tom Hamilton had warned Browne about the moral hazard inherent in the devolved business unit structure a decade earlier, and the regional presidents had been seen as a way of countering the incentive for managers to put short-term profits before safety. But now BP was dispensing with the oversight role of the regional presidents without appearing to give much thought as to how it would ensure BULs actually managed safety and environmental risks.

If anyone saw the dangers inherent in this decision, they didn't raise it at the time. Quite the contrary, in fact: Browne was seen as a visionary who had reinvented the way an oil company should be structured and then redrawn the oil industry by kicking off a round of the largest corporate mergers ever seen. If anyone peered inside at the workings of the company – and few equity analysts or investors ever bothered to do so – and spotted the change, it was seen as a clever streamlining of an already excellent system.

This was certainly the judgement of the CEO's management bible of the time: *The Modern Firm*, written by Stanford scholar John Roberts and named *The Economist*'s best business book for 2004. Noting Browne's decision to allow his ambitious and ruthless managers to regulate their own safety compliance without devising effective targets, the book concluded: 'It is arguable that much of the value created in BP's acquisitions of Amoco and Arco in the late 1990s came from applying BP's superior management systems to the American firms' human and physical assets.'

For years after the takeovers, the accepted history on Browne and BP would continue to read like fan mail. BP had become a carefully constructed time bomb, whose ticking was drowned out by the roar of tributes to Browne.

2

Climate Change Profiteer

In 2004 I was appointed to lead Reuters' coverage of the oil industry in Europe, the Middle East and Africa. In addition to guiding a team of correspondents across the region who reported on dozens of oil companies, I would have hands-on responsibility for covering the region's largest publicly quoted oil company, BP. With John Browne the industry's only celebrity CEO, we were always keen to secure interviews.

My predecessor gave me a piece of advice: if ever I was covering an event at which Browne was speaking, I should be sure to stand near colleagues from Reuters TV. Browne could be tricky about whether he would condescend to stop for questions from press reporters but if you had a TV camera with you, the BP boss would be more amenable. It was sound advice: I secured my first interview thanks entirely to the presence of a Reuters cameraman.

The oil industry is notoriously publicity-shy. Lee Raymond, chief executive of Exxon from 1993 to 2005, defined a good day as one on which Exxon stayed out of the newspapers; until the last few months of his leadership, when he gave interviews challenging global warming and deriding renewable energy, Raymond went to great pains to avoid TV cameras.

By contrast, Browne's only concern about talking to the media was whether the outlet involved had sufficient audience to warrant his time. Each quarter, when BP unveiled its earnings, the company invited TV teams from the BBC, Reuters, Bloomberg, ITN, CNBC and others to its headquarters in St James's Square to interview the CEO. The news teams would set up their equipment in a row of rooms on the sixth floor and Browne would move down the line from one to the next. If a station could not get a camera crew out to St James's, Browne would often jump into his chauffeur-driven Mercedes and go to them. His love of the cameras was often dismissed by his critics in the press as a sign of an ego ballooned and corrupted by power. No doubt there was some truth in that, especially during the later years of his tenure. But his breaking of the oil industry taboo on publicity was both a well-thought-out and deliberate strategy.

From an early stage, Browne had shown a willingness to adopt a public persona that was out of step with industry practices. Long before entering the chief executive's suite, he regularly gave press conferences and spoke at industry events. By 1987, eight years before taking the top job, *The Times* had dubbed him a 'management superstar', an unusual compliment for an executive of his level. Such press coverage did little to hurt Browne's internal reputation as the rising star, yet it was small change compared to what he dreamt of achieving. Plenty of executives have built a reputation by skilful manipulation of the media. Browne was unusual in realising he could build a whole business out of it.

When, as newly appointed head of exploration and production, he established himself in his Blackfriars office in the early 1990s, he surrounded himself not only with engineers and geologists to help run exploration and production, but also with specially appointed public relations and government relations advisers. In a truly unique move for a division of a corporation, he even appointed an investor relations executive for BPX (as the division was also known). He realised that, in BPX's case, the appearance of success was even more important than actual success.

In business, the truth will always come out in the numbers, but in exploration, it takes a long time for this to happen. Browne planned a bold new exploration strategy – focusing on big fields, or elephants, as they were known in the industry vernacular – in total contrast to the policy pursued by former CEO Peter Walters, who had aspired to a large collection of small fields. Browne was confident his plan would turn BP's fortunes around. There was just one problem: the time between making an oil find and establishing the full extent of the underground reserves is usually a period of years. Bringing a field on line can take even longer – a decade or more in the case of big projects – and in 1989, a lot of people didn't think BP would be around in five years, let alone the ten years it could take to prove Browne's strategy had worked. The company was pumping more oil each year than it was finding, which effectively meant that it was gradually going out of business. Short of a miracle, BP was surely on the verge of being taken over.

Browne decided the solution lay in perception. He was adamant that BP would, in time, become the industry's leading explorer. Until then, it would simply have to look the part.

Breathing hope into the market

In 1991, BP struck oil in tropical forests in the high plains north-east of Bogota, Colombia. The oil field it discovered was separated from the sea – and so from export markets – by a mountain range, and also bordered territory controlled by the Marxist FARC guerrillas, complications that had prompted BP to consider selling out. Now it was very glad it had not done so: Cusiana, as BP named the field, was clearly world-class.

It was the first success of Browne's new exploration strategy. If he could not quite assert it as proof that BP's declining reserves were finally on the up – it would be years before anyone knew how much oil was in the ground – he could certainly use Cusiana to bolster the company's image.

Since oil companies' share prices rise and fall on discoveries and dry holes, financial regulators impose strict rules of disclosure. Companies are not allowed to publish reserves estimates until they conduct a series of complicated tests that are supposed to give overwhelming certainty to the figures disclosed. But in the inevitable clamour for information that follows a new find, companies' investor relations teams often give analysts guidance that can point them in the right direction. The analysts then publish research notes in which they give field size estimates, but no source for their information. Everyone knows what is going on, and regulators, especially in the UK, are usually so helpful as to look the other way.

It was still early days in Colombia so BP could not disclose reserves estimates for Cusiana. Nonetheless, financial analysts began to describe it as a 2-billion-barrel field or better and some investors thought it could contain as much as 10 billion barrels.[1] At first, things went well and additional drilling over the next two years allowed BP to issue reserves estimates of 1.5 billion barrels. Discoveries by other oil companies operating in the country supported a growing perception that Colombia would become a major oil province.

Then things slowed. Appraisal drilling proved the boundaries of Cusiana did not extend as far as BP had hoped. If the company wanted to keep its shares moving higher, protecting it from takeover and paving the way for new acquisitions, it needed to find another big oil field – or, at the very least, another big story.

Browne decided he would need the right sort of person to drive the Colombian exploration venture forward. Who better than someone he had spent two years moulding?

By late 1992, Tony Hayward was coming to the end of his stint as a turtle. He was dispatched with his wife and two-year-old son to Colombia as head of exploration, the number two position in the country. The young local team immediately took to the import from England, partly thanks to his down-to-earth nature – Hayward would regularly join colleagues for beers and dancing at Bogota's most popular restaurant and nightclub – and partly thanks to his infectious enthusiasm. It was his first big job, and he was keen to distinguish himself. While BP was slashing jobs elsewhere, the Bogota office continued to grow: in the space of two years, the drilling fleet in the east of the country grew from just one rig to thirteen. Hayward would spend long nights in his Bogota office watching the fax machine as it churned out well logs from the new rigs, praying for signs of new discoveries.

But Hayward's work was not merely technical: he was also expected to be a cheerleader for the project, a role he took on with gusto. In 1993, reports began to emerge about another major prospect BP was exploring in Colombia. Hayward told reporters the company was eyeing two structures in a new exploration block named Piedemonte. He didn't say how much oil he thought the block might contain, but he professed great optimism.[2] Analysts predicted that Piedemonte could have twice the reserves of Cusiana.[3] As usual, it was unclear what the basis of these estimates was. It would be another two years before BP actually drilled Piedemonte. When it did, in the summer of 1995, it was quick to tell the world of its findings: two new fields, the Pauto and Florena, with estimated combined reserves of 750 million barrels of oil and 5 trillion cubic feet of gas.[4] Since this was the initial, official figure, investors naturally expected the final figure to be much higher, potentially the 3-billion-barrel-plus find that analysts had earlier mooted.

Hayward had delivered the goods. On the face of it, Pauto–Florena looked like another world-class find. Except that it wasn't. In time, BP would cut its estimates on Pauto and Florena to under a tenth of the initial figure.[5]

I am not suggesting that BP deliberately lied in its statement or its briefings to analysts or reporters. Indeed, the estimates may well have been the best the geologists could do with the information they had. But BP's claims were plain wrong, and unusual, in a number of respects. For one thing, the estimates were based on a single well drilled into each structure, a limited basis on which to offer detailed reserves estimates. Secondly, only one of the wells had been flow tested, the method by which a well's size is usually determined. It's a pretty simple principle: a high flow rate suggests a big field while a trickle suggests you might want to look someplace else.

The big international oil companies and the respectable independent explorers operated under a rule of 'under-promise and over-deliver'. It is the only way to maintain investor confidence over the long term. A combination of this principle and reporting regulations means companies usually start by issuing low reserves estimates for a new find, which are slowly raised as further drilling and flow tests are done. Based on all this, BP's release looked to industry experts like the kind of cheap trick that small, fly-by-night explorers on London's Alternative Investment Market, AIM, sometimes tried, when they wanted to ramp their shares up ahead of going back to investors for cash to fund additional drilling.

After Colombia, Hayward moved with his wife, his son and a newborn daughter next door to Venezuela, where he was made country head. Here again, his role appeared to be as much about making a big noise as about finding oil.

When the Venezuelan government announced a new oil licensing round – auctions in which oil companies compete for drilling licences – Hayward decided BP would use it to establish the company's reputation as a rising force in the country. He personally pored over seismic data and selected the licence he wanted most, dismissing challenges from others. Then he secured permission from London to bet big.

Venezuela had only recently reopened its oil industry to foreign investment and, in an attempt to appear transparent, the government decided to broadcast the sealed-bid auctions live on national television. So, when the time came, Hayward stood up in the Caracas conference hall and, in front of the cameras, handed over an envelope to the auction officiator. Inside was a $170 million commitment for the block he had selected. When officials compared it to the next highest bid, BP was declared the winner. Indeed, sources told me its bid was $80 million higher than the nearest offer. 'They simply blew us out of the water with their offer,' a representative for the loser said. The bid would be the highest made in the whole week-long licensing round.[6]

Hayward was bullish about the prospects, telling reporters, 'We feel that we got a pretty good piece of real estate there.'[7] But the block would never yield BP a drop of oil; in time, BP would write off the entire amount it invested. Nonetheless, for a time, it looked as if BP was about to repeat its apparent Colombian success in Venezuela.

Internally, the Colombian reserves and Venezuelan licence fiascos would earn Hayward the sobriquet 'Teflon Tony' for the way in which he managed to continue sailing upwards in spite of leaving messes behind him. 'He has always had an art of making things look good,' said one former senior colleague.

Hayward's critics were right, but what they didn't realise was that 'making things look good' was of major importance during his time in South America. 'Tony's role was to breathe hope back into the company,' one close aide of Browne said. Unfortunately, investors didn't realise that their hopes were also being managed.

'The more important part of Hayward being there was the talking up of Colombia, and the talking up of Venezuela, to the point where you could only be in the situation of over-promise and under-deliver,' explained Greg Bourne, BP's regional vice president for Latin America in 1997 and 1998. 'It probably served its purpose in terms of the stock market [by boosting BP's shares]. I'm not saying the stock market was being misled . . . But probably half a billion [dollars] was lost there, or even a billion was lost there, in terms of value.'[8]

By the time the reserves in Colombia were deflated and the Venezuela licence was shown to be worthless, investors, analysts and the financial press were too focused on the Amoco merger to notice – a merger that had, of course, been partly facilitated by the share rise BP received on the back of its perceived Latin American success.

This was just the beginning of a cycle in which BP spun investors and the public a rosy tale that, when everyone's attention had been shifted onto the next big story, would often as not turn out to be flawed. The positive news flow from Latin America during the 1990s gradually turned investors' perceptions of BP around. By the end of the decade, BP had dislodged Shell – in people's minds, if not in reality – as Big Oil's leading explorer.

The message was clear: people's perceptions could be changed dramatically and rapidly if one was bold enough, and if one spent enough money projecting the new image. John Browne would soon show there was almost no limit to how far one could push this principle.

Learning to love Big Oil

For decades, big international oil companies had been seen as the unacceptable face of capitalism. They supported dictators and undermined democratically elected governments. They despoiled the planet and dislodged indigenous people who were unfortunate enough to find themselves sitting on top of oil reservoirs. Big Oil was accused of manipulating oil prices in order to rip off motorists and producer nations. Big Oil was the perennial bad guy, the subject of a million conspiracy theories and the template for almost as many silver-screen villains.

The oil majors avoided the media spotlight because, any time it shone in their direction, it illuminated behaviour that brought them embarrassment and political attacks. They resigned themselves to their poor image. As long as not too much noise was made, they would always get by: demand for their product was not a matter of taste. One might avoid veal because it's cruel to calves or blood diamonds because they sustain conflicts in West Africa. Shunning oil wasn't really an option.

Negative news coverage sometimes hit business, possibly by prompting the removal of tax breaks or curbs on unethical international investment, as happened with Sudan and Burma. But politicians were slow to take harsh action. They appeared receptive to oil companies' claims that various crimes and harmful activities – environmental accidents, bribery, support for South Africa's apartheid government – were being moderated over time.

Conspiracy theorists believed this was because politicians, especially in the US, were in Big Oil's pockets. The majors were big donors but the bigger reason for the high level of political tolerance of the oil industry came down to fears about security of energy supply. The industry's argument was: if you tax us more or limit our ability to invest, you'll get less oil – and what you do get will be much more expensive. Politicians on both sides of the Atlantic remembered the queues at filling stations in the 1970s, and knew such scenes were not a recipe for electoral success.

By 1990, the bad reputation Big Oil had suffered was a growing problem in terms of recruitment, with the brightest young graduates increasingly disinclined to go into an industry which was seen as dirty, ethically challenged and lacking in innovation. Nonetheless, the oil industry felt it could probably solve that problem the way it did most things – throw money at it. Consequently, the oil industry entered the last decade of the twentieth century feeling unloved but secure.

Then something came along that shook the industry to its core: global warming.

For decades, scientists had warned that rising carbon dioxide emissions were warming the planet. By the late 1980s, political leaders had started to listen. In 1988, the United Nations formed the Intergovernmental Panel on Climate Change, to investigate and make recommendations in relation to global warming. Over the following years, the panel produced reams of evidence that the world had gradually been getting hotter. They identified the main reason for this as the burning of fossil fuels, namely petroleum, which was the largest single source of CO_2 emissions – 43 per cent in total.

By the early 1990s, Big Oil was really feeling the heat. For decades, oil companies had been able to explain their way out of their sins. Now it looked as if their greatest sin was their very existence. 'Up until climate change came on the agenda we could feel absolutely satisfied we produced a product that allowed economies to function and grow. It was absolutely noble. There were occasional environmental impacts but for the first time you had an argument that in burning our product, you were doing something essentially harmful. It was a bit like becoming a tobacco company,' said a former CEO of one oil major with a record of supporting despotic regimes across Africa.

The industry's initial response was denial. It challenged the science of climate change head-on and indirectly by funding think tanks and studies that sought to disprove the phenomenon. The US oil companies had an illustrious history in opposing political measures that might be injurious to their interests, and were initially confident they could do likewise with respect to climate change.

Their European counterparts were not so certain. They knew they could expect less support from their political leaders, and even doubted whether Washington would really remain inactive for ever. Mark Moody-Stuart, head of Shell at the time, feared that a catastrophic meteorological event – an exceptionally damaging hurricane, perhaps – might occur and be blamed on climate change. This in turn could prompt draconian political measures that would jeopardise the industry's very business model. He knew the bury-the-head-in-the-sand approach of Exxon, Chevron and Conoco would only hold for so long. John Browne also realised the oil industry could only halt action on climate change for so long, even in the United States, where the price of gasoline was as much an electoral issue as jobs or national security.

The question was what exactly to do? As was his nature, Browne examined the issue in forensic detail, consuming vast reams of data and speaking to many scientific experts and strategic advisers. It appears that he divined that, handled correctly, global warming could be turned from an existential threat into the oil industry's biggest opportunity in decades.

In 1997, Browne delivered a speech at his alma mater, Stanford University. The media and representatives of environmental groups were invited because Browne planned to make headlines. In the sweltering heat of an outdoor amphitheatre he acknowledged that the burning of fossil fuels was, most likely, warming the planet. He said that governments needed to start acting, and soon. BP, Browne added, was ready to do its part. It was prepared to contribute to climate change research, it would invest in renewable energy sources and it was prepared to subject itself to government measures that would penalise CO_2 emissions.[9]

It was a momentous occasion that duly generated masses of news coverage around the world. A Big Oil company had suddenly stopped denying climate change, apparently subjugating its own business interests to the greater interests of mankind. Browne called it good business. Climate change threatened mankind; ergo, it threatened BP's long-term survival. And while businesses usually tried to shift the cost of tackling broader social ills onto other parties, such as the taxpayer – the economic term was to 'externalise costs' – Browne said BP wanted to fulfil its obligations to society rather than be seen as a 'free rider'.[10]

The speech was followed by others on the same topic delivered in every corner of the world. Over the coming years, Browne would become a regular speaker on climate change at business events and on television.

BP began to spend heavily on advertising to emphasise its green credentials, and hired marketing professionals from consumer products firms such as Coca-Cola to reposition the company. Those who came on board were stunned by Browne's marketing savvy, unusual for an executive whose career was in producing and selling a commodity.

This drive culminated in the 'Beyond Petroleum' rebranding launched in 2000, a $250 million campaign that played up BP's minor investments in renewable energy and which won cabinets-full of awards for BP's ad agency, WPP. The rebrand also generated a new logo for BP: the Helios, a green and yellow sun that most people thought was supposed to be a flower.

It all seems crass and obviously false now, after the Gulf of Mexico oil spill and a raft of other BP disasters in the intervening years. Yet Browne's reinvention of BP's image held the world in thrall for almost a decade. His public comments won him a seat on the board of environmental group Conservation International, an invitation to smoke cigars and chat emissions with the Gubernator, Arnold Schwarzenegger, in his smoking tent at the California State Capitol, meetings with President Clinton at the White House, and a big spread in *Vanity Fair*'s famous 'Green Issue', which described the CEO as a 'committed environmentalist'.

Within a few years, Browne had transformed the perception of BP from climate villain to climate hero. His salesmanship was such that he even seemed to have convinced himself, to an extent, of BP's righteousness. But all the evidence pointed to a less philanthropic reality: BP was not, as Browne had suggested, offering to take pain today in order to avoid a systemic threat later on. On the contrary: BP didn't want pain today *or* tomorrow. And the greening of the company would ensure it didn't have to. It would help convince governments not to impose legislation that was harmful to its short- or long-term interests, and indeed to pass laws that would help BP sell more of its products, and at a higher price.

For all his talk of burden-sharing, Browne had set a course for BP to become the world's biggest climate change profiteer.

Moulding the debate

By the late 1990s, the vast majority of climate change scientists were unanimous in the view that the world needed to take urgent and dramatic action to avoid a climate catastrophe. In comments typical of the time, Klaus Töpfer, executive director of the United Nations' Environment Programme, said in 1998, 'We cannot afford to wait several years for the Kyoto Protocol to enter into force before making significant emissions cuts.'[11]

Browne argued that his conversion to the cause of climate change was inspired by conversations with such experts. However, in his speech in Stanford, he had warned that 'dramatic, sudden' action that 'sought, at a stroke, drastically to restrict carbon emissions' would be 'wrong'. Indeed, he argued that such action was not only unnecessary and impractical, but also 'discriminatory', because it would drive up energy costs and make it harder for undeveloped countries to emerge from poverty. If this rebuttal of scientific opinion disguised as an acceptance was brazen, BP's next trick was even more daring.

Logic generally dictates that success is best achieved by tackling the biggest problems first. Since petroleum was the largest single source of CO_2 emissions, one might expect that measures to tackle rising emissions would seek to curb consumption of this fuel source. Potential measures might include legislation switching large numbers of consumers from diesel and gasoline engines to electric vehicles.

The self-evident nature of this logic was the reason that oil companies had fought so long to deny climate change. Many people in the industry were aware that climate change denial was increasingly sounding as credible as claiming the earth was flat. The problem was that, in acknowledging a problem, one acknowledges the need to take action to solve it, and the obvious means of addressing climate change involved harming the oil industry. It was an apparently insoluble conundrum.

Then, in the mid-1990s, some in the industry hit a solution: find another villain. Browne may not have been the first to devise the plan but he certainly was the one who executed it the most successfully. The replacement villain was not mentioned in the Stanford speech, although Browne took the opportunity to dismiss the suggestion that oil was entirely to blame: 'Only a fraction of the total emissions come from the transportation sector,' he declared. True, oil was a fraction, but it was the biggest fraction. Now, he gradually became more specific in where the greatest problem lay and where action should be directed: 'Oil used in the transportation sector is only one source of emissions. Coal, power stations and industrial energy consumption are even greater sources,' he told an audience in Boston in 1998, in what could be described as a sleight of mathematics that compared a single element (oil) with several others combined.

Subtly, but deliberately, Browne refined his message, and soon the message became clear: coal was a bigger problem than oil. It was a dirty fuel that emitted dust particles, noxious, acid-rain-causing gases, and also much more CO_2 per joule of energy released than other fossil fuel. Browne pointed out – and gradually his rivals chimed in – that coal was especially more CO_2-polluting than natural gas. And the fortunate thing was that, when it came to heating homes and driving power stations, coal could so easily be substituted by gas. Coincidentally, the big oil companies had shed almost all their coal interests by the early 1990s. More importantly, they were awash in gas, having expanded their interests in this previously unloved hydrocarbon after being evicted from the biggest oil-producing countries in the Middle East in the 1970s.

Natural gas, like oil, is typically found in underground reservoirs; indeed, the two are often found together. But it is much harder to make money from gas than from oil, because it is more expensive to bring to market. Iran, for instance, has vast quantities of gas but no means to export it; one would have to spend tens of billions of dollars building pipelines across unstable countries, or construct multibillion-dollar terminals where the gas could be frozen and squeezed into liquid for export in liquefied natural gas carriers. By comparison, exporting Iranian crude simply involved pitching up in a tanker at one of Iran's ports. The difficulty in making money from gas was such that, for many years and in many places, a company that struck gas would simply cap the well and move on. Where gas and oil were found in the same well, the gas would be burned off on the rig and the oil exported to market.

But with oil investment opportunities harder to come by in the 1980s and 1990s, oil companies increasingly found themselves having to pursue gas projects. When Browne was made head of BP's upstream division, gas accounted for around 10 per cent of BP's total hydrocarbon production. In 1997, it was almost 20 per cent. By 2007, almost 40 per cent of BP's output was gas.[12] Demand for the commodity was growing but the sheer abundance of the stuff led to low prices and weak margins. The real challenge was to grow the market.

During the 1980s, oil companies had tried to encourage governments to support the expansion of gas use by highlighting its clean-burning properties: when combusted, natural gas produces only CO_2 and steam. This marketing policy achieved some success and, in the UK, helped natural gas to dislodge coal as a major home-heating fuel source. But coal remained the largest source for power generation in the world's biggest economies. This was partly a matter of tradition – the coal-fired power stations had been built decades earlier – but also a result of policy. Countries like the United States and Germany offered coal producers tax breaks and guaranteed markets, because coal provided an indigenous source of energy and created jobs.

When climate change hit the headlines, the oil industry suddenly had a powerful weapon with which to attack such incentives for coal. Browne realised that if gas could replace coal as a power generation source, the economics of the gas industry could be fundamentally changed for the better. 'His intention was certainly to build a gas market, because they had plenty of gas and no market,' said David Weaver, former managing director of BP's gas, power and renewables in Northern Europe.[13]

Browne was not the only oil executive who appeared to discover the business opportunities presented by climate change. Indeed, he was not really the first. A week before his rousing speech at Stanford, Shell had published its annual report. It contained a small section that acknowledged that global warming was, most likely, a reality. Indeed, Moody-Stuart would later enter a heated exchange of correspondence with Browne when the former Shell CEO spotted a BP advertisement on the London Underground claiming that BP was the first oil company to recognise climate change. (The exchange ended inconclusively with Browne accepting that Shell had been the first to publish an acknowledgement, but claiming that BP had been the first to acknowledge it internally.)

Why was Shell the first to publicly acknowledge global warming? They cited similar reasons to Browne – wanting to do the right thing, ensure the long-term health of the world economy, etc. – but tellingly, Shell was an even bigger player in the gas market than BP. Of course, Shell was also much more conservative, which precluded it from taking the forceful position Browne did. But oil executives fell in one by one behind Browne, and today even Exxon, after years of denial, is happy to acknowledge climate change and urge action. Of course, Exxon's business has also changed and in 2011 the industry leader is producing more gas than oil. To this day, when an oil industry executive talks about climate change, it is almost always to sell more gas. It is the reason why, when gas prices drop, such as in 2008, oil executives discuss the subject with greater frequency and intensity. Of course, they are always cautious not to overtly state their case.

Except when they slip up, as Tony Hayward did – and not for the first or last time – in early 2010. In a speech to a think tank in Washington, the then-BP CEO urged greater use of gas: 'Gas offers the greatest potential to achieve the largest CO_2 reductions – at the lowest cost and in the shortest time,' he said, echoing comments regularly made by oil executives. The claim was arguable from a scientific perspective and did not appear too self-serving. However, he then did what he often did when in the spotlight: he took it too far. For once, he made it abundantly clear who BP thought was the real climate villain and what should be done about it. 'It seems extraordinary', he went on, 'that the US is still focused on building coal-fired power plants,' adding that the coal industry had been 'disproportionately favoured' in a climate bill the US Congress passed in 2009.[14]

The coal industry and labour unions hit back hard. The United Mine Workers of America said Hayward had advocated 'putting an end to coal, the jobs that go with it and the retiree pensions and health care that coal pays for'.[15] The union called for a boycott of BP products. Hayward was forced to backtrack. He claimed that his remarks had been misinterpreted. Apparently, he had not meant to disparage coal at all.[16]

Emissions trading

After Browne's Stanford speech, BP, and then rivals such as Shell and Chevron, began pushing governments to adopt measures forcing businesses to cut their CO_2 emissions. BP and Shell's preferred option was an emissions trading scheme, or ETS, also known as 'cap-and-trade'. There are many forms of ETS but all are based on the principle that CO_2 emitters should 'pay to pollute'. The logic is that, if one has to pay to pollute, one will pollute less. An ETS allows parties to buy and sell permits that give them the right to pollute. Since the petroleum industry is the largest single source of emissions, oil companies arguing for the 'pay to pollute' principle sounded like turkeys voting for Christmas. Many environmentalists and politicians took the support BP and others offered for emissions reduction plans as a clear sign of good intentions.

BP led the charge, setting up internal teams to study how the company would operate under an emissions trading scheme and also advising governments on how they might implement an ETS. 'BP did a fair amount of work on emissions trading internally, as well as how it might be designed externally, so they put some effort in on the policy side that I think was valuable,' Eileen Claussen of the Pew Center on Global Climate Change said in 2011.[17]

The Pew Center is one of the leading international climate change groups and boasts a board that includes the former United Nations Environment Programme director Klaus Töpfer and Theodore Roosevelt IV, the former president's great-grandson. Eileen Claussen was one of the opinion leaders Browne sought to influence, offering her a special invitation to the Stanford speech. What Claussen and political leaders did not appreciate was that BP didn't see ETS as a way for Big Oil to share the burden of tackling climate change. Rather, the company saw emissions trading as a potent mechanism for accelerating the shift from coal to gas.

Europe was the first to implement an emissions trading scheme and the model it adopted was the one designed by BP. Indeed, given the absence of any indirect or direct penalty on the oil industry, the EU's ETS could scarcely hide its heritage. The petroleum industry's responsibility for CO_2 emissions is largely indirect, through consumers' use of its fuels. Yet despite the key contribution of cars, trucks, planes and boats to total CO_2 emissions, the transport sector was exempted from the EU emissions scheme. This wasn't a hard sell. European governments already taxed fuel sales heavily; adding new taxes on top would have been politically unpopular across Europe.

But the oil industry is also a major CO_2 emitter in its own right: oil platforms emit masses of CO_2 because they run large turbines to power their facilities, because they flare some of the gas they produce to keep production levels stable, and because some reservoirs release CO_2. Refineries are also major emitters of CO_2 since they burn up to 15 per cent of the fuel they process as part of the crude distillation process. Nonetheless, thanks to lobbying by BP and its peers, refineries and oil platforms were effectively exempted from the scheme.[18]

On the surface, this cunning sidestep looks remarkable. Up close, it's even better. When designing its ETS, the EU chose a gradualist approach, whereby companies were given permits to emit 92 per cent of their expected CO_2 emissions for free. This left them with a choice: they could either produce less CO_2 – by limiting operations – or they could purchase the additional credits they needed. Conversely, if a company found itself with excess credits, perhaps because it became more efficient and emitted less CO_2 than it did in the past, this company could sell on its excess credits. From the perspective of a power producer, this was a clear incentive to replace a coal station with gas. Since a gas station emits half the amount of CO_2 of a similar-sized coal plant, a coal generator who switched to gas would find itself with much more CO_2 credits than it needed (42 per cent of the original total emissions). These spare credits could then be sold on the open market.

Of course, since they were exempt, the oil companies got all the credits they needed for free. This proved an opportunity to make another killing. Refiners burn a portion of the oil they process to run their operations. This burnt oil represents one of the refiner's biggest costs and is the main source of its direct CO_2 emissions. In an effort to cuts costs during the lean years of the 1990s, refiners installed more energy-efficient equipment to reduce their own oil use, which also pushed down CO_2 emissions. This meant that when the EU imposed its ETS in 2005, Shell and BP found themselves with excess credits that they could then sell on to the market at a profit.[19]

Regulation is a matter of swings and roundabouts for businesses. One incurs direct, short-term costs but the upside is the long-term health of one's industry. The casual observer might have concluded that such give and take was what Browne was referring to when he said he didn't expect a 'free ride' for BP when it came to tackling climate change. Yet that was exactly what BP and its rivals lobbied for when the EU was drawing up its ETS. They pushed the same old buttons they always did, saying the inclusion of refineries and oil rigs in the EU's ETS would make EU energy infrastructure uncompetitive, lead to the closure of refineries and platforms and thereby jeopardise Europe's security of energy supply. As usual, the argument was successful. In every major respect, actions to tackle climate change benefited BP and its peers. In the UK, a combination of Margaret Thatcher's reduction of support for the coal-mining industry and a desire to cut CO_2 emissions helped push the percentage of the nation's power generated by coal from 70 per cent in 1987 down to 37 per cent by 2005.[20] Conversely, gas's share of the mix rose from 1 per cent to 35 per cent over the same period. If the same turnaround could be repeated in big coal burners like the US and Germany, the expansion in the global gas market would generate hundreds of billions of dollars for the oil and gas industry.

Browne said in Stanford in 1997 that he wanted BP to 'contribute to the public policy debate in search of the wider global answers to the problem' of global warming. What he showed was that, by contributing very skilfully, he could not only espouse policies that cut emissions but also turn them into a massive bonanza.

Green energy

Browne knew that BP could not establish its green credentials – and thereby influence government policy – merely by talking about climate change. The company needed to do something concrete to convince people it was genuinely working towards the low-carbon economy.

One solution was to do what environmentalists had long urged oil companies to do: invest in renewable energy. The only problem was that renewable energy didn't make any money. BP's board knew becoming a big player in green energy would involve committing vast amounts of capital for little or no return.

The eventual plan was something of a halfway house: BP would invest a little bit of money in renewable energy and then spend a lot of money publicising it.

BP had formed a tiny solar operation in the early 1980s,[21] and would regularly cite this as proof that renewable energy had always been in its DNA. The truth was that the company's entry into solar power was part of an industry-wide diversification drive prompted by fears about future access to oil reserves, following OPEC's evictions in the 1970s. As part of this same drive, BP also entered the animal feed business and metals mining, while rivals bought into newspapers, biotech and forestry. Indeed, around the time BP dipped its toe in the solar market, Exxon was doing the same, although Lee Raymond would never have the audacity to claim this proved he had green genes.

In the decade and a half since BP's solar venture was formed, it had lain almost dormant, receiving only the smallest trickle of investment. Then, in 1997, as Browne prepared to deliver his Stanford speech, BP agreed to spend $20 million on a bankrupt solar power manufacturing plant outside San Francisco. A year later, the Amoco takeover gave BP a half-share of a joint venture with Enron, called Solarex, which manufactured solar panels. Shortly after BP completed the Amoco acquisition, it bought out Enron's stake. By integrating all these stakes under the banner of BP Solar, BP became the biggest global manufacturer of solar panels. The fact that one achieved such a position for an investment of a few hundred million dollars was a sign of the embryonic status of the solar industry at the time.

On the face of it, solar was a strange choice for an oil company, even one that wanted to enter the renewable energy business. BP's core business was to extract and process energy. It did this by dint of its geological and project management expertise. The solar panel business had absolutely no overlap with BP's competencies. First, solar was not an energy business but a manufacturing business. It was firmly in the electrical goods camp – a game of product design and product differentiation that oil companies knew nothing about. Secondly, oil companies weren't really customer-facing. They had fuel retail operations but these made little money and were gradually abandoned from the late 1990s onwards (although the companies who bought their stations often retained the oil majors' branding). Thirdly, solar deviated from BP's core skills because it was small. BP marketed itself to resource holders as a company with a special expertise in big things. Its total investments in solar were so small by the end of 1999 that they were less than the amount subsequently spent on rebranding the company as Beyond Petroleum.

In fact the only good thing, commercially, about solar was that it looked good. Despite concerns about whether it actually cut CO_2 emissions – and BP acknowledged that it took six years for a solar panel to generate as much energy as was used in its construction[22] – it suffered almost no public criticism. By contrast, biofuels were accused of either leading to deforestation or pushing up food prices by competing with crops for land. Windmills faced regular opposition for blighting the landscape and decapitating birds. Solar was simply the greenest of the green.

Environmentalists saw the paucity of BP's investment in green energy as a sign that the company's environmental message was simply 'green wash'. Analysts and investors saw BP's minimal financial outlay as a sign that profitable investment opportunities did not exist and that the business would always remain small. Some feared solar would always require more management attention than it was worth.

Browne countered criticism by saying he was examining other green energy areas and that he hoped to build a material business in one of them. The spiel went that BP wanted to be prepared in case the world shifted away from hydrocarbons. But all the studies conducted by the industry, including by BP researchers, said it was unlikely that wind and solar would account for more than a few per cent of the total energy mix by 2030, while oil and gas would continue to account for around 60 per cent. This was clearly not a rational basis for diverting investments from hydrocarbons to renewables. Perhaps that was why, when it came to committing capital, Browne demurred. David Weaver, managing director of BP's gas, power and renewables in Northern Europe, had often pitched green energy plans to no avail. When he finally convinced the company to allow him to build a wind farm – BP's first ever, coming five years after Browne's famous speech – it was only because he said he could secure outside project financing. Similarly, when it came to solar, pitches for cash proved very difficult. Browne said he wanted BP to remain the biggest global player in solar but his refusal to properly invest saw BP Solar overtaken by others such as First Solar, which created a profitable, multibillion-dollar business just as BP Solar floundered.

It wasn't simply a case of reluctance to commit capital. The fact that solar was more about burnishing BP's reputation than making money sometimes precluded it from doing the latter. When BP opened its solar panel manufacturing plant outside San Francisco in 1998, Browne was there to welcome the likes of Vice President Al Gore and *Cheers* star Ted Danson.[23] The high-profile event emphasised BP's environmentally friendly credentials and, argued Browne, showed that BP was taking a leadership role in a fast-growing market.

Within a few years, however, it became clear that solar panels were becoming a low-margin business and that the key to survival was to minimise manufacturing costs. This meant shifting production out of high-cost places like California. But California was high visibility, and that was what really mattered. 'What we didn't do was redirect manufacturing capability to China quickly enough,' said a former boss of the solar unit. 'It was an error in business strategy. It was not taking the difficult decisions to close plants in the US . . . There was a real concern about the reputational damage of doing that in a country which is extremely important for BP [as an oil and gas market].'

BP slowly entered the wind industry and then, following legislation in the US and Europe requiring a percentage of all fuel sold at forecourts to be plant-derived, biofuels. But investments remained relatively tiny: less than 1 per cent of total capital investment between 1999 and 2005. The disparity between green energy's prominence in company advertising and its weight in investment expenditure became an increasing embarrassment to BP. It was a dilemma shared by rivals such as Shell and Chevron, which by the mid-2000s were also seeking to be seen as environmentally concerned energy giants.

One way to avoid this embarrassment would have been to increase spending on renewables. Another might have been to stop running advertisements that focused on green energy. The industry chose a third option: companies stopped publicising their spending, or lack of, on renewable energy. Then they got very creative.

In 2005, BP formed BP Alternative Energy, ostensibly to meet the world's growing demand for low-carbon energy. The division would invest up to $8 billion over ten years in greener energy sources such as wind, solar and hydrogen, compared to around $1.5 billion spent on renewables in the previous decade. But as was often the case with BP, closer examination showed a less ecologically impressive picture. The apparent quintupling of BP's commitment to green energy was actually being largely generated by integrating natural gas-fired power generation operations into the renewables portfolio. Indeed, natural gas projects accounted for at least half of the 'Alternative Energy' portfolio.[24]

If this seemed like an obvious sleight of hand, it wasn't so obvious as to be spotted by many people who reported on the new division. To this day, BP's $8 billion commitment to Alternative Energy is frequently reported, in even respected journals and books, as a promise to invest $8 billion in renewable energy.

That said, in terms of creative green energy accounting, BP would soon be outdone. As part of Chevron's 'Will you join us?' campaign, which some industry executives interpreted as the Chevron CEO's attempt to steal John Browne's green crown, the California-based company made an impressive claim: 'Between 2002 and 2006,' company literature announced in 2007, 'Chevron spent approximately $2 billion on alternative and renewable energy technologies in such diverse areas as geothermal, hydrogen, biofuels, advanced batteries, wind and solar.' All of a sudden, it appeared that the oil business had an unsung hero in renewable energy – by this stage, Shell and BP had both invested only around $1 billion in wind, solar and biofuels. It was an especially impressive achievement since, up until the end of 2004, according to that year's Corporate Responsibility report, Chevron had only invested $60 million in renewable energy projects. It appeared that 2005 and 2006 had seen it pump almost $1 billion, or around 10 per cent of its capital expenditure, into wind, solar, geothermal, etc. each year.

Except that it hadn't. In fact, the company only spent around $100 million, mainly on a new geothermal facility, over the two years. The $2 billion figure was almost entirely accounted for by the 'alternative' tag that Chevron took (as did BP) to include gas-fired power generation and energy efficiency measures. The latter could include anything from installing a more efficient crude distillation unit at a refinery to replacing an old television set in a smoking room on an oil rig with a flat-screen TV.[25]

Boardroom unease

BP's board had given its approval to Browne's ground-breaking speech at Stanford acknowledging climate change. Yet, for many directors, it was a grudging support. 'It was a necessary piece of external communication,' said one executive director from the time. The board knew it had to do something in response to climate change and that denial was no longer the most productive strategy.

BP's directors and senior managers also saw that the company enjoyed ancillary benefits from the socially responsible reputation it developed:

for one thing, a smoother ride for its takeover of Amoco than might otherwise have been expected. The US is notoriously tetchy about large foreign takeovers, especially in strategic industries like energy. Congress didn't block the deal, and BP's directors felt the green image probably helped. Moreover, a greener BP found recruitment so much easier. While rivals struggled to attract good graduates, BP saw a steep rise in applications from Ivy League universities over the course of the 1990s.

But the sheer extent to which BP rebranded itself as an environmental champion troubled many senior figures at the company. The board worried that BP's image could demoralise employees in the core business by making them feel top management was not focused on their activities (as two directors told the author). The green message also confused investors.[26] Goldman Sachs came in and told Browne the image was weighing on the company's share price. The investment bank said some investors feared the company was going to direct cash away from high-margin oil production into lower-margin renewable energy. And BP's board also worried the Beyond Petroleum message would make BP look dishonest. A company selling itself on the tag line 'Beyond Petroleum', when petroleum remained, and was expected to remain, the company's principal business, was a contradiction no more lost on the directors than on the environmentalists who derided it as 'green wash'.

Yet what worried some people most of all, including BP's head of media relations Roddy Kennedy, was that, in claiming to have such elevated principles, BP was setting itself up for a fall. If something went badly wrong, and BP was perceived to have fallen short of its socially responsible claims, the kicking it would receive from the media and policymakers would be all the harder. The image-makers behind Beyond Petroleum scoffed at the naysayers who rained on their green parade. 'Roddy thinks his job is to keep BP out of the press,' WPP chief executive Martin Sorrell joked to a friend.

In the end, there would only be one way to know who was right.

3

There's No Such Thing as Santa Claus

In early 2005, as John Browne approached a decade leading BP, his reputation was riding higher than ever. In the UK, he had achieved the status of national business treasure, while internationally he found himself feted by presidents, oligarchs, kings and despots. To investors, he was simply the man who kept BP cruising along.

There had been setbacks. In 2002, BP was forced to cut its oil and gas output targets three times in eight weeks.[1] It had to have been a gross embarrassment for the CEO and a direct result of the overly ambitious goals – known in management speak as stretch targets – he had outlined two years earlier. Investors were disappointed but the reality was that, even after the reduction in forecasts, BP was growing at least as fast as its rivals. Indeed, Shell's production was shrinking.

In any case, Browne made up for this disappointment a year later by cutting a deal with a group of Russian oligarchs to form TNK-BP. The joint venture would be Russia's third-largest oil producer and add a million barrels per day to BP's production. Browne convinced Vladimir Putin to endorse the deal, the only one of its kind before the president turned against foreign investment in Russia's oil industry.

In 2004, BP's shares rose faster than those of almost all its rivals. Perhaps more tellingly, the stock also enjoyed what analysts describe as a 'management premium': it traded on a higher price–earnings multiple than companies like Shell, France's Total or Chevron. In buying BP shares, investors were prepared to accept a lower short-term return, in the expectation that BP would outperform over the long term.

Indeed, BP's reputation for efficient management was so high that the British government leaned on it for guidance. High-flying young executives were sent on secondments to government departments.[2] Tony Blair appointed BP's chief of staff, Byron Grote, to advise on improving productivity and efficiency in the public sector,[3] and Browne gave Blair lectures on motivating managers and target-setting.[4] And when the government got it wrong, Browne did not hesitate to admonish it. In January 2005, he spoke in Davos about how Blair's creation of 'pseudo markets' within the NHS – new targets under which patients were seen more quickly, but for shorter and less thorough examinations – was damaging patient care. Narrow target-setting was, in Browne's view, leading to undesired outcomes.[5]

Perhaps it was a form of arrogance begotten of ten years in office, but suddenly he had developed an acute ability to spot flaws in others to which he was blind in himself. If Browne were to look at his own business in early 2005, he would have seen countless examples of how narrow target-setting and weak oversight had created perverse incentives for BP's managers. But so long as the production, cost and profit figures looked good, he and BP's investors didn't appear to ask too many questions.

Rogue traders

One department in which Browne showed a particularly stunning ability to overlook evidence of perverse behaviour was the oil trading unit. Oil trading was an area BP had entered almost by accident: having been thrown out of the Middle East in the 1970s, it found itself short of crude to feed its refineries and charged Bryan Sanderson, a rising star who would one day head BP's chemicals division, to set up a team to source crude. Sanderson realised that by buying and selling oil, BP could hide the fact that it was short of the stuff, a weakness that rivals might seek to exploit.

Over time, however, BP saw that buying and selling crude could be highly lucrative in its own right. The company's strong market position – built on the ownership of oil fields and refineries, and on its superior access to industry knowledge and marketing networks – gave it an edge over the other traders, many of whom were smaller independents with few physical assets. Moreover, since BP had a good understanding of likely demand and supply over time, it could turn a profit betting on its hunches. Indeed, for a little extra risk, BP could even determine what the price would be in the future. This was achieved by executing what was known as a 'squeeze'.

Squeezes have gone on from time immemorial. The idea is that if someone corners a market in a product by buying up all the supply, people who need the product will have to come to them and buy at any price. Of course, cornering a market is difficult. One has to store a lot of product and it's a highly visible activity, inviting unwanted attention onto the squeezer. The development of derivatives markets allowed a more sophisticated and subtle squeeze – one BP learnt to exploit with ruthless efficiency.

The modern squeeze usually operates like a mirror image, with the squeezer deliberately setting out to lose money on the products he or she buys. It may seem counterintuitive that a trader would seek to manipulate a market for such a purpose, and indeed the new squeeze is not intended as a money-losing activity. The logic behind it lies in the fact that oil derivatives markets had grown to around ten times the size of the physical market (known confusingly as the 'cash' market). Canny companies are able to use activity in the physical market to influence prices in the much larger derivatives market.

As an oil broker in the 1990s, I had a few opportunities to observe BP's team in action in the European gasoline market. Dealings involving the sizeable Amsterdam–Rotterdam–Antwerp market (known in trading circles as ARA) were particularly interesting to watch, since the prices at which barges of gasoline were bought and sold at ARA were used to price North West Europe gasoline derivatives. Perhaps inevitably, the gasoline-barge market was a regular subject of squeezes by BP and others. We brokers acted as intermediaries between traders, and could always tell a squeeze was under way when a trader said he or she was eager to buy gasoline barges, but then only agreed to ever-increasing prices. For example, if a trader running a squeeze bought a cargo at $260 per tonne, he or she would then seek to buy the next one at an even higher price. Naturally this stood out, since buyers usually want to buy things more cheaply. The ever-increasing levels of the trades would duly be recorded and the official prices of gasoline derivatives would duly rise. The barges, full of thousands of tonnes of gasoline, might later be sold at a loss since they were purchased at inflated prices, but it didn't matter. If the squeezer bought 100,000 tonnes of physical gasoline, it invariably meant he or she owned ten times that volume's worth in derivatives and the real aim was to drive up the price of these.

BP was famous for such activities in European crude and oil products markets throughout the 1980s and 1990s. Physical oil markets and the 'over the counter' market in derivatives – those entered into directly between companies, rather than via an exchange – were not overseen by government regulators, which means the behaviour was not illegal. Exchanges such as the New York Mercantile Exchange (NYMEX) and London's International Petroleum Exchange (IPE) were regulated, and traders knew not to engage in sharp practices here. But for some the temptation was too great. In 1997 and 1998, crude trades conducted on the IPE by BP caught the eye of regulators. The IPE ended up fining BP £125,000 while the financial regulator, the Securities and Futures Authority, fined the individuals involved.[6]

BP's strong position in Alaska also allowed it to use some of the tricks honed in Europe on the US West Coast. In the mid-1990s, BP traders began exporting Alaskan crude to Japan, even though it could be sold at a higher price to refiners in California. US regulators would later determine that this represented a deliberate attempt to squeeze the West Coast crude market.[7]

Such fines should have been a red flag to BP's management that the pressure its traders were under to generate profits was prompting them to sail too close to the wind. But the lure of profit was too much for BP, especially after it integrated Amoco, whose position in the Midwest offered an even bigger platform from which to play, or indeed manipulate, the markets. Indeed, in the years following the Amoco purchase, BP found itself in constant run-ins with energy regulators. In 2000, the company's US energy trading team saw its rivals at Enron make a killing on the California power crisis (Enron was later even blamed for causing the power shortages) and decided it wanted a piece of the action. The Commodity Futures Trading Commission (CFTC) subsequently accused BP of engaging in 'wash trading', the execution of phoney trades that sought to manipulate prices. BP paid $100,000 to settle the charges and the traders involved paid $55,000. Questionable crude oil trades in 2001 and 2002 brought a $2.5 million fine from NYMEX.[8] BP's trading of gasoline futures contracts in 2002 prompted another CFTC probe.[9]

BP's senior management were not blind to what was going on. The discovery of the California crude squeeze nearly scuppered Browne's takeover of Arco and meant that regulators only gave their approval to the acquisition after Browne agreed to sell big chunks of Arco's assets. This considerably weakened the economics of the deal. Over the same period, its rivals Exxon, Chevron and Shell did not have to report any similar trading probes or fines,[10] so it is unlikely that management was taking the view that legal infractions were like injuries on an oil rig: unwelcome but unavoidable. Meanwhile, a former head of trading even sat on BP's main board, so top management could not claim they lacked the available expertise needed to understand exactly what the company's traders were up to.

More likely, the deafness to the persistent trading abuses reflected the lure of the short-term profits on offer. And what profits: $3 billion in 2005.[11] At one point, BP's trading desks made four times as much money as all its refineries and forecourts combined. Fortunately for BP, energy trading was such an arcane area that its infractions garnered little media coverage. How could a mainstream newspaper hope to explain 'wash trades' and 'squeezes' in a 250-word story?

But problems of an altogether more visible nature were bubbling away in another area of BP's portfolio – problems that highlighted that, when it came to choosing between short-term targets and the long-term health of the business, Browne's structure of independent business units effectively left managers with no choice at all.

Soiled wilderness

The Prudhoe Bay oil field, the largest in the United States, was discovered in 1968. To move the crude to market, a pipeline was built from Alaska's North Slope to the ice-free southern harbour of Valdez, on Prince William Sound. The Trans-Alaska Pipeline System (TAPS) cost $8 billion and required a workforce of 70,000 to build. At the time, it was the most expensive privately financed construction project in history.[12] Nonetheless, when the engineers sat down and drew up their development plans, they made some basic mistakes. These errors would rapidly take the lustre off what should have been a showcase for the industry's abilities.

A little over a decade after oil began flowing south from Prudhoe Bay to Valdez, serious corrosion was detected in the pipeline: engineers had underestimated how much water would be produced with the crude. By the time John Browne began to re-engineer BP's upstream division – the unit responsible for Alaska – the facilities at Prudhoe Bay and the TAPS were in need of serious unforeseen investment.

At first, the sheer remoteness of the Alaskan oil installations kept the problems a secret. Then workers began to leak information about the deterioration of the infrastructure to the press. When the Exxon Valdez tanker ran aground in March 1989, causing what was at that time the country's worst ever oil spill, Alaska's government was spurred into action and told BP and the other oil companies to put their houses in order. The solution to the problem was obvious: spend more money on improving the facilities. But this was an unpalatable proposition for a company that was undergoing a savage restructuring aimed at slashing costs. This conundrum provided the context for one of BP's most controversial business decisions since its support for the coup that returned the Shah to Iran in 1953.

The Alaskan oil workers who leaked information about corroded pipelines, both the TAPS and BP's own, usually did so via an intermediary named Chuck Hamel, a former oil trader who had had a falling-out with Alyeska, the consortium of companies that operated the TAPS, over a cargo of crude laced with water. Hamel would pass the information to the media or environmental groups. Alyeska considered him as a major thorn in its side.

Since BP was the largest shareholder within the Alyeska consortium, it usually appointed the man in charge of the TAPS. In the early 1990s, this was James Hermiller, a long-time BP executive, who decided that the pipeline had a communications problem rather than a corrosion problem.

If only, Hermiller reasoned, the leaks to Hamel and his associates could be stemmed, all the political pressure on Alyeska would go away. He hired the Wackenhut investigation agency to make it happen.[13] Wackenhut's agents secretly taped Hamel's phone calls, intercepted his mail, stole items from his garbage, and went through his credit card records. The agency even hired attractive female operatives to try to entice him into compromising situations and into betraying his sources.[14] The plan was uncovered and led to a major scandal. An Alaska judge slammed Alyeska for using tactics reminiscent of 'Nazi Germany',[15] Hermiller was replaced[16] and Congress set up an investigation. Millions of dollars were paid out in civil settlements. Hamel's award provided him with the necessary financial security to devote himself entirely to leaking details of problems at Alyeska and BP's other Alaskan operations. There was much work for him to do.[17] It seemed that BP's executives were so focused on the narrow targets they had been set that they could not see the wood for the trees. Costs had to be cut almost regardless of the consequences.

Former turtle John Manzoni was one such executive swept up in Browne's mantra. Having been sent to Stanford to do a Master's in business, he was then dispatched to Alaska to manage BP's North Slope operations. It was his opportunity to put what he had learnt from Browne into practice and he was keen to use his first post-turtle job to showcase his management potential. He did indeed make a name for himself in Alaska, although not the one he might have chosen. His zeal for cutting costs was so absolute that, for over a decade after he had moved on, Manzoni was remembered as 'the man who brought the union onto the North Slope by cancelling free ice cream'. This was a little unfair. There were a number of reasons why workers at Prudhoe Bay sought the protection of organised labour: the gradual erosion of the strong terms that had drawn them to the inhospitable terrain in the first place, reduced staffing levels which put a higher burden on remaining workers, and bullying of workers who reported problems such as pipeline corrosion.[18] The removal of a minor perk was unlikely to have been a significant factor in the unionisation of the workforce. Nonetheless, the ice cream story was taken as a sign of the bluntness and absoluteness with which Manzoni would work towards driving down costs, and under him, the pipelines continued to deteriorate.

Alaska was not a problem created by Manzoni. The unintended flaw in Browne's corporate structure[19] was that managers like Manzoni could only fix such problems at the expense of their career. By the early 2000s, the deferred maintenance in Alaska, the rusty valves and holed pipelines, had led to field shutdowns that cost BP tens of thousands of barrels per day in lost production. When the repairs finally began to be executed, they were more expensive than if they had been undertaken earlier. But by that stage Manzoni had moved on. The next guy would have to deal with the problem.

BP was an organisation that consistently provided incentives for staff that in effect acted against the broader, long-term objectives of the company – and even of the industry.

The election of George W. Bush raised oil industry hopes that, after decades of waiting, the Arctic National Wildlife Refuge would be opened to drilling. When Bush included revenues from ANWR lease sales in his budget, it seemed a positive decision could be just months away. But continued spills at BP sites during 2001[20] and an internal report detailing 'large and growing' maintenance backlogs on fire and gas detection systems and pressure-safety valves, among other problems at BP's Alaska facilities,[21] helped kill any hope of ANWR drilling.

Refining problems

BP reached the end of the twentieth century in rude financial health, but every division was showing the scars of long-term underinvestment and cost-cutting. It was perhaps inevitable that the least favoured and least profitable part of the business would suffer most. Refining – boiling crude down into usable fuels such as diesel and gasoline – had a bad name throughout BP because of a long run of losses. Around the turn of the century, the board regularly discussed whether the company should exit the crude processing business altogether. Other companies faced a similar quandary. In 2000, Paul Skinner, Shell's head of refining whom BP would later court to be its chairman, put it succinctly: 'The blunt fact is that most refining investments over the past twenty years have not delivered an adequate return.'[22]

But none of BP's rivals had quite the same laser-like focus on capital efficiency. A lot of capital was tied up in refining and the return was not commensurate with the earnings a similar amount invested in the upstream could generate. This meant that refinery managers found it extremely difficult to convince senior bosses to spend money in refining operations – and BP's decentralised management and tough performance contracts made arguing too strongly seem a potentially career-limiting option.

After the merger with Amoco in 1998, Browne delivered a demand for 25 per cent cost cuts across the group. In this, he was not altogether out of step with his rivals: the late 1990s round of mega-mergers was premised on cost-cutting. Bringing together two companies achieved big savings by eliminating overlaps in areas such as information technology and equipment design. When these integration synergies had been squeezed out, however, the industry went back to business as usual. But Browne had other ideas and derided his peers for their lack of ambition. 'The fashionable view seems to be that we've done all that – and that there is little further potential to reduce costs,' he declared at the 2000 World Petroleum Congress, the biggest bash in the oil industry calendar. 'I don't agree. I think the potential for further gains from cost productivity is extensive.'

The result of this continued squeezing was inevitable: maintenance was put off, and obsolete or substandard equipment was not replaced.

In 2000, BP had a wake-up call. It suffered three accidents over 13 days at its Grangemouth refinery in Scotland, including a major fire that Britain's safety regulator, the Health and Safety Executive (HSE), said had endangered the lives of workers and civilians. Its review of the slip-ups blamed the incidents on poorly maintained equipment and cost-cutting. What the regulator didn't know was that the refinery's former manager, Paul Maslin, had actually asked for more money to improve the plant but been turned down.[23] When he persisted, he had been demoted.

The HSE also deemed BP's practice of measuring safety to be flawed. BP argued that its refineries were becoming demonstrably safer, since the number of reported injuries was falling. The first problem with this was that the reported figures were probably wrong: workers at some BP facilities would later complain that they were discouraged from reporting their injuries by managers who knew rising rates would reflect badly on them. But even ignoring this, the fundamental practice of using injury rates had, by 2000, been discredited as an indicator of overall workplace safety, since they mainly capture minor accidents – sprained wrists – and ignore the risks that cause major disasters, such as potentially lethal gas leaks. The minor incidents that show up in injury rates are mainly the result of lapses in personal safety, while accidents that cause multiple deaths are usually the result of process safety flaws: unsafe equipment and procedures. Process safety comes down to the level of effort on the part of the employer.

The regulator told BP it needed to look beyond what went wrong when workers spilled hot coffee or failed to hold the hand rails on stairs. The company needed to invest in making its facilities safer and to provide training so that people knew how to operate safely. Browne said BP would learn the lessons from Grangemouth,[24] and it should have been easy for him to make good on this promise. After all, BP was, experts agreed, the ultimate 'learning enterprise'.[25] Its peer groups were supposed to spread best practice throughout the organisation like a virus. But the peer groups weren't really designed with safety in mind. When managers are measured on financial performance only, it is inevitable that any knowledge-sharing forum created for them focuses on sharing ideas that can boost profits. Safety was not something the peer groups got bogged down in.

BP accepted the HSE's report on Grangemouth and professed an epiphany in the area of process safety. But it was like a pledge of abstinence on a hung-over Sunday morning that's quickly repealed in the afternoon. Rather than improve, BP actually went in the opposite direction.

In the early 1990s, BP had had regional presidents who were responsible for ensuring that BULs – business unit leaders, such as refinery managers – complied with environmental and safety policies. At the end of the decade, mere months before the Grangemouth incident, Browne had decided this supervision was becoming a drag on performance, and devolved responsibility for ensuring compliance on such matters to the BULs themselves.[26] It effectively created self-regulation on health and safety. Browne, the man focused on hard financial targets, appeared to take a more laissez-faire approach when it came to setting health, safety and environmental objectives, which were usually expressed as vague goals rather than firm targets. Managers would be told not to damage the environment, to gradually make their workplaces safer and to be good citizens in their locality. The system of self-regulation was so wishy-washy that many managers would later tell BP internal investigators they had not even realised they had been given the responsibility for monitoring their own performance in these areas.[27]

Where clear targets were devised, they accounted for only a small part of the overall performance contract – often around 10 per cent of the bonus managers received for meeting or surpassing their performance obligations.

And what did these targets measure? If there was any company in the industry who should have known what to look out for and measure when it came to safety, one might have thought, after Grangemouth, that BP was it. So did it incorporate process safety measures – such as the number of hydrocarbon leaks or fires at a facility, or even the number of near misses – into managers' performance contracts? No. BP continued to measure safety in terms of injury rates, the kind of personal-safety metrics that the HSE blamed for the complacency that nearly killed many workers at Grangemouth.

In a context where managers' bonuses and promotion prospects hinged on meeting Browne's demands for ongoing cost reductions, the absence of process safety targets created a moral hazard for managers, encouraging them to underinvest in their facilities. If anyone doubted the power of such incentives, or the lengths to which managers would go to avoid spending money on their facilities, events at BP's Carson refinery near Long Beach should have settled the matter.

In 2002, the California body charged with regulating air quality became suspicious that the Carson plant might not, as it should, be reporting all the accidental releases of fumes from its petroleum storage tanks. Releases did occasionally occur at refineries and penalties were imposed to ensure these leaks were minimised. At Carson, minimal noxious releases were vital because the refinery was located near residential neighbourhoods.

Staff shortages in the 1990s forced the Californian air quality agency, the Air Quality Management District (AQMD), to give oil companies the responsibility to report their own violations. By 2002, the agency had become worried about BP's filings. Somehow the company's level of violations had dropped sharply despite virtually no reported repairs to its facilities. Officials sought access to the plant to inspect the tanks, as was their right, but BP refused to let them in. The company said health and safety rules precluded it from allowing anyone near the tanks without the proper safety equipment, such as breathing masks. The officials did not have this gear with them, but BP did. The plant had the equipment on site for just such an eventuality, yet declined to share it with the AQMD officials. The regulators later returned with the Los Angeles County Sheriffs in a fleet of SUVs. When they inspected the tanks, the inspectors found more than 80 per cent of them were in breach of regulations. The agency sued BP for $319 million, and accused the company of falsifying reports.[28]

If BP's bosses were shocked by events at Carson, they didn't show it. The refinery's operations manager, Colin Reid, was later promoted to manager of the much larger Whiting refinery.[29]

In 2003, BP appointed a new executive responsible for health and safety across the whole company. Greg Coleman had joined BP through its Canadian division, and was another former Browne turtle. Shortly after he took up his post, Coleman went on a tour of the company's US refining plants. When he returned, his verdict left little room for interpretation: 'It frightened the shit out of me,' he told a friend. The equipment at facilities including Toledo, Whiting and Texas City, the largest of BP's refineries, was dilapidated and the staff were not following basic safety procedures.

Coleman shared his fears with Mike Hoffman, the group vice president in charge of refining. Ideally, BP would have shut down at least parts of the facilities while the necessary safety changes were made, but everyone knew this was never going to be approved. Coleman suggested a compromise whereby a full list of the necessary repairs was drawn up and a programme instigated to tackle the work over a period of a few years. This plan would naturally involve considerable additional money being spent, and production levels would be hit.

Hoffman agreed the facilities were badly in need of investment and the two men approached Hoffman's boss, the overall head of refining and fuel marketing, and made a pitch for the money. Unfortunately for them, the man they were pitching to was John Manzoni.

Manzoni had been made head of refining and marketing in 2002. Like all ambitious executives, he always looked two steps ahead. A job had to be seen in terms of the job it led to afterwards. In 2002, Manzoni was doubtless looking two steps ahead to the group CEO job. And he knew he wasn't the only one: his friend and rival Tony Hayward had recently been named company treasurer, a role held by Browne in the 1980s and which ensured exposure to all the financial expertise – debt raising, foreign exchange risk management, representing the company to bankers and investors – that a future CEO would need. Manzoni felt the pressure. His own new role was not the best springboard to the CEO suite – by the turn of the millennium it was clear that BP's future was in the upstream – but it was still his best chance to make a pitch for the top job. And given that Browne and the board constantly agonised over the amount of capital tied up in the refining sector, and clearly didn't want that side of the business to grow, there was only one way to create value: squeeze it. Manzoni had quickly identified two priorities: the shape of the refining portfolio and competitiveness. He decided that many of BP's refineries were located in low-growth markets like Europe and so, where possible, facilities would be sold. Elsewhere, costs would be cut.

Consequently, Hoffman and Coleman's pitch for more investment was rejected. Hoffman would later try again to argue for cash to tackle the decrepit facilities he managed but each time, Manzoni said no. 'He got tired of Manzoni saying, "No, no, no",' one former executive vice president said of Hoffman.

Even though everyone knew it was a measure of limited value, Manzoni pointed to the falling reported injury statistics as a sign that everything was all right. It wasn't simply that Manzoni had never worked in a refinery before, or that, sitting in London, he was too far away from the rotting US refineries to see the process safety risks building up. When he was deposed by lawyers for families of the victims, he said that, though he could not recall it, he must have been provided with the 2002 report that warned of 'serious concerns about the potential for a major site incident' at Texas City. He admitted that Hoffman had told him he wanted to spend more money to upgrade the facilities.[30] Perhaps Manzoni didn't act to improve process safety because he had no incentive to do so – in fact, he had every incentive not to. The perverse effect of his incentives was reflected in the fact that, while he ignored warnings from the heads of safety and refining about the risk of a major catastrophe, he was overzealous in seeking to stamp out minor safety failings. When he noticed that vehicle accidents were boosting injury statistics at US refineries, he announced his plan to go over to Whiting and fire anyone he found driving around the refinery without wearing a seat belt. People laughed when they heard this, until they realised he was serious. Someone had to take him aside and explain that this would likely involve sacking half the workers and sparking a labour dispute.

Nonetheless, on the financial front, Manzoni's efforts quickly began to show dividends. BP's 2003 annual report commended him for achieving 'substantial gains in productivity throughout the segment'.

Texas City

Texas City was BP's largest refinery, a former Amoco plant built in 1934 and capable of processing 450,000 barrels of crude per day. But even by the unattractive standards of its peers in BP's refining portfolio, it was, by 2005, an ugly beast. Visitors noticed broken windows, uncut grass, peeling paintwork, roofs in need of repair and corroded pipework. During the lean years in the 1990s, Amoco had underinvested in the plant to the extent that, by the time of the 1998 takeover, it was already in need of attention. What it got instead was further cost cuts.

In retrospect, BP executives said the refinery suffered especially badly from the cuts because of a clash between the Amoco and BP cultures. BP executives were used to negotiating their performance contracts with their superiors. They challenged every target put to them. Since the contract would be enforced mercilessly, no one wanted to agree to a target they did not believe they could hit. Better to agree a target below what you could achieve and then exceed it. Amoco managers, on the other hand, were used to a more top-down culture. Senior managers would consider the capabilities of their equipment and people and set targets – be it in cost reductions or production – accordingly. Lower-level managers were expected to accept these goals without question. If, during the year, it became clear that the target was unrealistic, the senior managers would adjust it. Amoco managers were not used to haggling with their bosses.

Even after the merger, Texas City's staff and managers remained almost exclusively former Amoco employees. Well into the 2000s, people at the plant wore hard hats and overalls with the Amoco logo. So when Browne issued his 25 per cent cost cut target in 1999, the ex-Amoco managers approached it just as they would have any target from their former bosses. Consequently they ended up slashing costs even more rigorously than most. 'The senior management in Texas City didn't understand intimately how you got stuff done in BP,' said Colin Maclean, the former North Sea hand and Browne confidant who had led the integration of the BP and Amoco refining operations after the merger. 'The money was not procured effectively, and a lot of stuff that should have been done was not done.' The consequence was a total end to practical training at the plant. Only the most basic maintenance was conducted, and outdated and unsafe equipment, which rivals were phasing out, remained in use. The lack of investment hit morale and contributed to sloppy work practices, thus adding to the dangers.

The degradation of the facility did not go unnoticed. Even before the chilling assessment of new health and safety chief Greg Coleman, several reports had flagged the dangers. In 2002, having learnt enough about BP's culture to realise that senior bosses were more likely to listen to external consultants than their own managers, the Texas City leadership hired consultants AT Kearny to conduct a review of the plant. The review said cost reductions implemented in the wake of the merger were contributing to a decline of infrastructure that would require significant investment to correct. It noted the large number of fires that occurred at the plant and said these posed a risk of a major accident at the site.

A year later, an internal safety audit noted that 'the condition of the infrastructure and assets at Texas City is poor', and complained of a 'checkbook mentality' when it came to safety. In 2004, another external review noted underinvestment at the plant, while another internal audit noticed that the backlog of repairs was rising. The various managers at Texas City passed these reports up the line but were muted in their demands for more cash. 'The guys who ran the refinery. They were afraid for their jobs, there was a culture of fear,' said a former executive vice president.

In the middle of 2004, Texas City got a new manager, Don Parus. Shortly after he started, two employees were scalded to death while removing a valve from a hot-water line. Earlier that year, a contractor had been killed in a fall inside a gasoline-making unit. Deaths were rare in the refining industry, so three in a year stood out. Parus went back and looked at the history of the plant and discovered deaths were not at all unusual at Texas City: it had been killing workers at a rate of almost one a year for over a decade. With fires breaking out at the plant 50 to 100 times a year, it was surprising that there were not more deaths. (Ironically, smoking was banned in most areas because of the risk of starting fires.)

Parus began forcefully lobbying senior management for more cash. When John Manzoni visited the refinery in July 2004, Parus showed him a presentation that included excerpts from some of the reports conducted in previous years. There was no response. Parus followed this up with other pleas to his superiors and even visited London to make his case, each time without success. By the end of the year, he felt like his constant harping was going to cost him his job. Nonetheless, he made one last stab for more cash. He commissioned consultants Telos to conduct yet another review and told them to not hold back. They didn't. Telos said managers felt they had to compromise on safety to meet production goals. Employees reported 'feeling blamed when they had gotten hurt' and 'an exceptional degree of fear of catastrophic incidents at Texas City'. Outside contractors at the refinery had seen equipment degradation at the plant that they had not seen at any other places they had worked.

Parus passed his report up the line. He might have expected action. The refinery had just posted $1 billion in profit for 2004 – its best year ever – so for once there was plenty of cash about. What he actually received was a 'challenge' to cut another 25 per cent off his budget.

On 23 March 2005, workers were restarting a gasoline unit, a raffinate splitter, which had undergone maintenance. They accidentally overfilled the unit, something that should have been detected by a level indicator and an alarm – but neither was working properly. The liquid overflowed into pipes connected to another unit, known as a blowdown stack – basically a tall barrel-like structure with what looked like a 100-foot chimney on top – which had been designed decades earlier to disperse fumes that emanated from the raffinate splitter. The blowdown stack had the obvious flaw that if enough vapours escaped, they could settle on the ground and ignite. This risk was well known, and it had in fact been industry practice for over a decade to phase out blowdown stacks and replace them with flares for burning the vapours safely. Rather than replace its blowdown stack, however, BP had added an additional line into it, increasing the potential for disaster.[31]

As the raffinate splitter continued to overflow, the blowdown stack filled with flammable liquids. Control room operators – some of whom had been working 12-hour shifts for up to 30 days straight due to staffing cuts – were unaware of this because the alarm inside the stack was also broken. One hundred and thirty feet away, a group of contractors were taking a break in some wood-framed trailers. Trailers were supposed to be at least 350 feet away from such gasoline units, although they were permitted to be closer if a detailed review had proven it would be safe. Such a review might have allowed blast-proof steel trailers within the boundary. BP deemed them too expensive.

Shortly after 1 p.m., workers saw a geyser-like eruption from the top of the blowdown stack. It lasted about a minute.

A pickup truck was near the blowdown stack, in yet another breach of procedures, and bystanders noticed its engine begin to run faster as it sucked in vapours from the evaporating liquid. Moments later, the fumes ignited and triggered an explosion that was heard miles away. Three quarters of a mile away, homeowners suffered smashed windows and other property damage. Within the refinery complex, hundreds of people were injured. Heavy men were knocked to their feet by the shock waves, some breaking bones and others receiving deep gashes to their limbs. Workers sitting within trailers as much as 500 feet away from the gasoline unit suffered injuries.

But these were the lucky ones. Those in the wood-framed trailers 130 feet away from the stack did not stand a chance. Fifteen workers were killed, mostly immediately, including a husband and wife, James and Linda Rowe. The bodies of the dead were so badly damaged that investigators would later have to rely on dental records and DNA to identify them. In Linda Rowe's case, the task was complicated by the fact that her head had been taken off by a piece of flying debris.

Dozens of others suffered up to 90 per cent burns. The blast shattered bones, peeled skin and flesh from victims' bodies and left limbs hanging off, while scalding smoke seared the insides of lungs. The plant resembled a bomb site. Fires raged and men screamed in agony. Those workers who could move scrambled to the fence surrounding the refinery to try to get away, but the barbed wire on top stopped them. One worker backed a truck up against the fence and threw his overalls over the barbed wire so his colleagues could escape.

One hundred and seventy men and women were taken to hospital in ambulances, while hundreds of others suffered less serious injuries. Many of them would never recover from their wounds. Some would not walk again and all would bear psychological scars.[32]

After years of warnings, the disaster that had been predicted had finally happened. As the truth about Texas City's recent past became known, people asked how senior management could have failed to act on the warnings they received. Some blamed BP's 'aggressive' corporate culture, a rather abstract concept. The real answer was simpler and more concrete. Managers did not act to prevent Texas City because every incentive and potential penalty they faced told them not to.

A PR 'incident'

It would be wrong to say that the first thing senior BP managers thought of when they heard about the Texas City blast was the potential fallout for the company's reputation. But that was certainly near the top of the list. Three hours after the explosion, Ross Pillari, the head of BP America, exchanged emails with his PR adviser. He was reassured that, though there would likely be damaging headlines the following day, an early Easter break would come to BP's rescue. 'Expect a lot of follow-up coverage tomorrow, but I believe it will essentially go away due to the holiday weekend,' wrote the press adviser in an email later uncovered by the victims' lawyers.

Almost immediately, a plan for dealing with the media and public was formed. This was partly superficial. Managers were briefed on how to address the media. For example, they should refer to the tragedy as an 'incident', and most certainly not as an 'explosion', 'blast', 'fatal tragedy', or anything else that might be accurately descriptive. BP's dogged use of this one word to describe the disaster, for years afterwards, would become a source of macabre amusement for reporters covering the tragedy.

Such tactics were insincere, yet largely harmless. What happened next was infinitely more alarming. In the immediate aftermath, no one knew what had caused the Texas City blast, but BP immediately decided that it was not the fault of the senior executives: BP as an institution was not to blame. Critics believed that the alternative – admitting responsibility on the part of senior management – meant it might as well set fire to the hundreds of millions of dollars it had spent cultivating its green, socially responsible image over the previous eight years. The company needed to shift blame, but at the same time not give the impression that this was what it was trying to do. The solution it appeared to arrive at was ingenious: a blameless mea culpa.

On arrival at Texas City, Browne immediately accepted responsibility for the 'incident' and said BP would fulfil all obligations to its injured workers and the families of the dead. It sounded nice and BP would go on to claim great credit for it, but the admission cost them nothing. The blast happened on BP's facility so it was undeniably the company's responsibility, while Browne must have known the law would ensure it fulfilled its financial obligations to the victims. One thing the acceptance of responsibility was not was an acceptance of guilt. Quite the contrary. When Browne held a press conference in Texas City Town Hall a day after the blast, he declared: 'There is no limit to the amount of activity that we've undertaken in Texas City to make it a very safe plant. And it is a very safe plant.'[33] Sitting behind a bank of microphones, he also denied any link between the blast and previous deaths at the plant.

It was an amazing set of assertions, not least because Browne had only arrived a few hours earlier and had received just a brief tour of the wrecked plant. Safety regulators and BP itself would later agree these claims were not even a little bit accurate. Perhaps that was why Browne for once looked flustered. He was dressed impeccably in his tailored suit, crisp white shirt, and a sombre dark-green tie. Yet his face had a defensive look that those who knew him did not recognise. Everything he had worked to create over the past decade and a half was suddenly at risk, and he knew it.

But such denials could only go so far. In Europe, industrial accidents are usually investigated over a period of years by commissions who are reluctant to criticise. The investigation into the 2005 Buncefield fuel depot explosion in Hertfordshire, for instance, Britain's biggest peacetime explosion, took over three years to complete and only hinted at blame on the part of site operator Total. It was five years before the French oil giant was fined. Similarly, the HSE's stinging report on BP's Grangemouth accidents took three years to write. In the US, however, a combination of swift and determined official investigations and aggressive trial lawyers ensure that people don't have to wait so long to know whom to blame for their pain. BP knew this and decided it wouldn't passively wait for the process to take its natural course, potentially to the company's detriment. It immediately conducted its own review.

Two months after the disaster, BP America boss Ross Pillari called a press conference in Houston and released the results of the company's preliminary review, which concluded that three factors played a role in the disaster: the placement of the trailers close to the gasoline unit, the use of a blowdown stack, and failures in operating procedures. Pillari admitted that no safety review had been conducted on the trailer placement but added, 'The investigation does not indicate it contributed to the incident.'[34] He suggested there was nothing unusual about locating a trailer so close to a gasoline unit and said the real problem was the unforeseeable ferocity of the blast. He omitted to say that the refinery's safety procedures precluded trailers being placed within 350 feet of dangerous equipment, without the completion of a safety review.

In terms of the use of a blowdown stack instead of a flare, Pillari accepted this played a role but said the alternative would only have 'reduced the severity of the incident'. He insisted the equipment was intrinsically safe. The 'primary' cause of the accident, he said, was 'surprising and deeply disturbing' mistakes made by the low-level workers who overfilled the raffinate splitter. Pillari did not mention that alarms that were supposed to have alerted workers to the overfill were not working. Six junior employees were fired immediately.

Pillari's conclusions would later be contradicted by safety regulators and by subsequent BP investigations. Nonetheless, BP began aggressively promoting its message. The company doubled its advertising spend[35] and told publications running BP ads that it wanted to be notified of any negative stories on BP that they planned to carry.[36] If the publications did not forewarn BP, advertising could be suspended. The company also boosted its government relations spend and launched a programme called 'Fabric of America' to emphasise its community contributions.[37]

At the time, communications professionals thought BP had hit a home run. 'On a PR basis,' concluded one website for public relations professionals, 'BP handled the catastrophe as a class act.'[38]

Blowback

Due to the time difference between Texas and London, I first heard about the refinery blast early in the morning on the day after it occurred. Given BP's reputation at the time, I did not immediately suspect that bad practices were to blame. This view soon changed when I received a call from a former BP worker with experience of its refining unit. The source said they feared the blast was the result of something deeper than simple human error, adding that BP's refining operation was a mess due to years of bad management and underinvestment, and that top management knew it. Every other year, BP would poll 100,000 of its employees to seek their views on a range of issues and to measure morale. Recent surveys had shown that 'basically people didn't want to be working in our refineries any more'. Management had been so concerned about it that a programme called 'Operation People' was introduced in an attempt to improve the situation, although the programme was largely cosmetic.

The source provided BP documents that proved the existence of morale problems and a programme to address them. When presented with this evidence, BP was hostile. Eventually, it was forced to accept the truth about the worker morale surveys, but it denied that this had any role in the disaster. It also claimed that subsequent surveys had shown a turnaround in morale, an assertion it would later drop. Indeed, the company's most senior safety executive for refining would later admit that its People Assurance Surveys still showed a 'very demoralised workforce' ahead of the blast.[39]

Our report was published on 24 March 2005[40] but not picked up by our media clients, who appeared, in post-9/11 fashion, to be more focused on the potential for terrorist involvement in the blast. Nonetheless, my Reuters colleagues in the US and reporters at other agencies gradually began to discover details of previous fires and fatalities at Texas City. These were widely reported. So when Pillari dumped the blame for the accident on low-level workers, he was not talking to an altogether naive audience. Indeed, it was pressure from the press that forced BP, days after Pillari's press conference, to withdraw its claim that the six sacked workers were the 'primary' and 'root' cause of the explosion.[41]

In the coming months, tales of earlier safety warnings and the decrepit state of the equipment emerged. The safety regulators provided interim reports and updates which, combined with material uncovered by trial lawyers, maintained a steady stream of bad news for BP. It was clear that the company needed a new strategy if it wanted to preserve its image. Browne hired former US Secretary of State James Baker to conduct a review that would tell BP – and, once published, the world – what really went wrong at Texas City. Baker's law firm, Baker Botts, was one of the biggest serving the energy industry, while his think tank, the Baker Institute, had previously invited John Manzoni to sit on a panel that was formulating energy policy recommendations. Browne figured Baker had the right blend of high reputation and sensitivity to the oil industry to help forge a new narrative on what happened at Texas City.

Baker's panel provided some embarrassments for BP. For one, he found that 'cost cutting was not carried out as a structured, managed, and measured process', which contradicted Pillari's earlier claim that 'the approach to reducing costs was well thought out and systematic'.[42] But he also found that managers had not deliberately cut safety-critical expenditure. Rather, the report concluded that BP had simply got its wires crossed. The company had mistakenly identified falling personal injury statistics as a sign that the overall safety of the plant was strong. It was an honest mistake, and not a sign of any structural weakness. Browne accepted the report and professed now to fully understand process safety, his second such epiphany in five years. (The final report of the US Chemical Safety Board, issued in March 2007, would not let the company off the hook and held cost-cutting directly to blame for the blast, although it didn't challenge the view that BP had ignored process safety too.)

Browne appointed a new safety guru and began giving lectures on the importance of process safety. It might have worked were it not for the fact that, as part of her settlement with BP, Linda and James Rowe's daughter, Eva, insisted that BP release vast numbers of internal documents that would normally have been sealed as part of such a settlement. These documents torpedoed any hope that the disaster might be rebranded as an honest accident.

The disaster also killed Manzoni's hopes of replacing Browne. The discovery of an email he had sent a colleague, in which he noted having had to interrupt a family holiday in Colorado to travel to Texas City 'at the cost of a precious day of my leave', was simply the icing on the cake. The email gave the impression that he didn't care, which was unfair. Shortly after the blast, a colleague told me she had walked into his office to find him distraught. Other colleagues would testify to the fact that Manzoni 'took it personally'.

But falling on one's sword, or even being pushed onto it, wasn't an option: it would have implied culpability on the part of the company. An internal probe recommended Manzoni's position be considered, and that everyone in the chain of command between him and Texas City be fired.[43] But none of them were. Manzoni, his deputy, Hoffman, and head of US refining Pat Gower still had to go out and bat for the team. For over two years, they would have to provide depositions and turn up for court hearings, to deny that safety had deliberately been sacrificed in the name of profit, or that anyone had ever warned them about the dangerous state of Texas City. The men remained on the company payroll, as did Don Parus, the refinery manager who, all agreed, was severely emotionally affected by the accident, until the lawsuits against the company were settled.

BP denied outright that the Texas City blast was a sign of any greater malaise in the company. But over the following year, further disasters raised further questions.

In July 2005, BP's *Thunder Horse*, the world's largest semi-submersible oil and gas platform, was evacuated in line with safety rules as Hurricane Dennis approached. When the workers returned in helicopters they found the $1 billion platform listing at a precarious angle. It soon transpired that valves had been installed incorrectly, causing water to enter the vessel's hull. Various other design and manufacturing failures were discovered as repairs were undertaken. The flaws were the direct result of BP's design, chosen in the hope of meeting Browne's overly ambitious production targets. Instead of building two smaller platforms, BP had plumped for one massive, groundbreaking platform, because it would allow the field to ramp up to full production more quickly.[44] The combination of an ambitious design, a rushed delivery and BP's eroded technical capability made problems almost inevitable.

Photographs of *Thunder Horse* at a 30-degree tilt made front pages around the world, further weighing upon BP's reputation. Yet worse was to come.

In March 2006, BP reported a 5,000-barrel spill from a leaking pipeline in Alaska. It was the largest oil spill ever to occur at Prudhoe Bay. Probes would show, and BP would be forced to admit, that cost cuts were to blame. In the wake of Texas City, the public's reaction was predictable. Then, just when it looked like BP's reputation couldn't sink any lower, the financial regulator, the CFTC, uncovered tape recordings that they said showed BP traders had manipulated the US propane market. It wasn't the biggest or highest-profile energy market, but the fact that propane was a fuel that poorer people relied upon made the attempts to ramp up prices seem all the more heartless.

Browne went on CNBC and professed outrage: 'Whatever the rights and wrongs of this case, these tape recordings show that a few people in BP broke our values and we don't like that.' He spoke with the shock that only someone totally ignorant of BP's long rap sheet on trading fines could claim. Internally, the scandals challenged the true believers' faith in the Browne revolution. 'It was like suddenly discovering that there's no such thing as Santa Claus,' said one former Browne turtle.

When another leak was discovered in Alaska in August 2006, forcing the shutdown of half of Prudhoe Bay, even investors began to lose patience.

BP's shares had hardly been dented by Texas City and the first Alaska spill. Indeed, in April 2006, analysts at Lehman Brothers calculated BP shares still traded on a premium to all peers, including Exxon. Dresdner Bank explained that 'BP continues to be comparatively highly rated and deserves to be.' A month later Citigroup commended BP's 'trusted management savvy'. The problem was that shareholders and analysts had previously believed what BP told them. Initial estimates of the cost of Texas City were $200–400 million.[45] The final bill was over $3 billion.[46] *Thunder Horse* repairs were predicted to cost $250 million and take a year to complete. The final bill was put at $1 billion and the work took three years.[47]

When Prudhoe Bay was shut down in August 2006, with a loss of hundreds of thousands of barrels per day in production, BP's shares lost their 'management premium'. Combined with the protracted repair of

Thunder Horse, analysts realised their earnings forecasts were too optimistic and cut them. The consequential cooling in BP's shares saw Shell overtake the company as the world's second-largest oil group by market capitalisation. It was clearly time for change.

Sunset for the Sun King

The mega-mergers of the late 1990s had propelled Browne to the top table of international business leaders. With his rising profile came an increasingly imperial manner. Everything around him was of the best: office furniture was designed by the queen's nephew, David Linley; he had two drivers so that he could be ferried around in his enormous black company Mercedes at any time of day or night; he travelled with an entourage of assistants, policy advisers and occasionally press advisers; he flew by BP jet for company business and first class for private journeys, until frustration at occasionally finding himself in a seat other than 1A (the ultimate status symbol in commercial travel) drove him to take out a personal subscription with NetJets.

For older directors who remembered chairman and CEO Peter Walters being driven around London in a Ford in the 1980s, it seemed excessive. Some feared that even BP's environmental message was increasingly being driven by Browne's desire for public adulation. Concerns also grew about how he related to people internally. When Browne's executive team presented to the board, Browne would, in the words of one director, 'top and tail' their presentations, providing an introduction and summation of their comments.

Directors noticed that some of the company's best people were walking out of the door. Peter Backhouse, formerly executive vice president, refining and marketing, had been seen as a possible Browne successor, until he left in 2000. Then Dick Olver, head of exploration and production, who had also been seen as a CEO candidate, departed to become chairman of arms maker BAE. There were others.

This left a coterie of former turtles beneath Browne. All were bright and able but also seen as yes-men, moulded in their leader's image and unlikely to challenge him. Indeed, board members knew from their own experience that challenge was something Browne did not welcome.

But there was one person at BP who did not hesitate to challenge Browne: Peter Sutherland, BP's chairman, and technically Browne's boss. Sutherland was a Jesuit-educated barrister who had been Ireland's youngest ever Attorney General. From here, he was appointed a European Union Commissioner for competition policy. This led to him being asked to lead the General Agreement on Tariffs and Trade (GATT), an international body trying to broker the biggest international trade deal ever. When GATT was replaced by the World Trade Organization, he was asked to lead this. When he finally quit public office, he was made chairman of Goldman Sachs International, the investment bank's overseas arm. This part-time role netted him a $100 million fortune when Goldman floated on the stock market. In 1997, he became BP's non-executive chairman.

In personal terms, Browne and Sutherland were polar opposites. Sutherland was overweight with heavy jowls, while Browne was short and slender. Sutherland, a former rugby player, liked to watch sport, while Browne favoured opera. The Irishman liked to crack jokes, sometimes rather irreverent ones, while Browne was not burdened with a sense of humour. (One colleague noted that Browne had never been heard telling a joke nor seen listening to one.) Sutherland was seen as the definition of gregarious, whereas Browne was on the record as saying he wasn't. Both men appreciated the finer things in life, such as good food and wine, but Sutherland clearly enjoyed them more.

Despite their obvious differences, the two men seemed to have struck up a good personal, as well as professional, friendship. They would smoke cigars together at company events, socialise together and even holiday together. But people close to the men saw a deliberateness about these actions. 'They did an awful lot of things together to appear to be of the same point of view and good friends,' said one US-based director, while a British captain of industry who was a friend of both men described the relationship as 'an acquaintanceship with mutual respect, up to a point', but added, 'In an ideal world for John, there would have been no chairman. John wanted to control every issue at BP. It was a constant source of tension between them, but they worked out an accommodation.'

As a non-executive, Sutherland was not supposed to get involved in the day-to-day management of the firm, but he was supposed to provide an independent layer of oversight above executive management. Standing back was quite easy since he didn't have a background in business, but his political experience made him highly attuned to matters of reputation and this sometimes led to clashes with Browne. He worried, for instance, that Browne's tie-up with the Russian billionaires could sully BP's image, and insisted no Russian representative be admitted to BP's board. He also expressed concerns about how it would look if BP pushed headlong into repressive China. Increasingly, he served as a counterpoint to the deal-mad Browne and when, from around 2002, Browne began to talk hypothetically to colleagues and advisers about possible benefits of merging with Shell, Sutherland made his scepticism of such a venture known. 'Peter saw it as his responsibility to the world to guard the company from John Browne,' said one non-executive director.

By the end of 2004, even Browne's peers at the top of other Big Oil companies were discussing the rift between the BP chairman and the CEO. But in time, the differences would go beyond pure business matters. When Browne's mother died in 2000, he was totally devastated. His devotion to her was not simply a matter of dutifulness. He was as reliant on her as she was on him for emotional support. He had even taken her with him to investor roadshows in America (much to the chagrin of the brokerages hosting the events, who had to find someone to take her shopping while Browne did his presentations).

At the same time, Browne's relationships had remained clandestine affairs while his mother was alive. After she died, he became more reckless. He started to surf websites for male escorts. In 2002, he met a 23-year-old Canadian called Jeff Chevalier through a site named Suited and Booted. Nine months into their relationship, Chevalier moved into Browne's Chelsea home. The couple concocted a story about having met while jogging in Battersea Park, which they decided would travel better than the truth. Ironically, the story prompted winks and nods among some friends, who suspected Chevalier might have been working as a prostitute in the park. Nonetheless, everyone was happy to accept what was at worst a white lie.

Sutherland had known about Browne's homosexuality before the CEO disclosed it to close colleagues. He called Browne into his office a few years into his tenure as chairman and told Browne that he should never hesitate to come out for fear of embarrassment to BP. Browne, friends said, was touched by the comment. Sutherland told him to bring Chevalier along to company events and invited the couple to stay at his house in La Zagaleta, one of Spain's most exclusive gated estates, located outside Marbella. 'It was schizophrenic,' one London-based board member said. 'One minute John was in the closet and the next Jeff was everywhere.'

Despite all the signals of support, however, Browne's colleagues and friends had privately not taken to the flighty Chevalier. Many thought the age gap a problem, especially since Chevalier seemed so prone to sulks. They tolerated him for Browne's sake but became increasingly concerned that the relationship would end badly.

And in spite of his early support, Sutherland too became concerned when he heard details of Browne's previous relationship failures. It might have been the start of the enlightened twenty-first century, but there was still reason to fear unflattering headlines regarding a captain of industry and male lovers decades his junior. When Browne mentioned in a 2001 interview with *Forbes* magazine that he collected photographs of flowers by Robert Mapplethorpe, who was also well known for his homoerotic photos of nude men, BP received letters and emails from homophobic readers across America, who said they would never buy BP gasoline again. There was also a commercial risk. Browne clearly wasn't comfortable with his sexuality being known to the wider public, which made him an obvious target, if not for blackmail, then for pressure of more subtle kinds. The risk was seen as especially keen in respect to BP's partners in TNK-BP. In Russia, it was standard practice for businesspeople to gather compromising material on partners and rivals which could, if necessary, be leaked to the media or authorities at a convenient juncture. A term had even been coined to refer to such material – 'kompromat' – and every oligarch worth his salt was expected to have a vault full of it. Indeed, Browne heard at one point that the oligarchs had been digging into his personal life and was much alarmed by it.

Sutherland shared his concerns about potential media or commercial embarrassment with fellow board members. 'It was an issue that Peter discussed quietly on a number of occasions,' said one non-executive director. 'He could feel that it was a potential problem he needed to make the board aware of and be protective about.'

The relationship between Sutherland and Browne deteriorated to the extent that even friendly gestures illustrated the disharmony. In December 2005, Sutherland invited Browne and Chevalier to his daughter's grand wedding at Dromoland Castle in the west of Ireland. Browne was one of the few business or political associates Sutherland invited. The vast majority of the 200 or so guests were family or personal friends. But despite this honour, the wedding was not a success for Browne. Chevalier had one of his sulks and retired early. The sight of Sutherland, the 18-stone former prop forward, throwing shapes on the dancefloor while his wilder Irish friends downed pints of Guinness proved less than entertaining for the opera-loving and Montrachet-drinking Browne. 'It was like a weekend from the theatre of the absurd,' he later told a friend. The semblance of unity was wearing very thin indeed.

Desperate measures

Standard practice at BP was for senior managers to step down when they reached 60 years of age. Shell had the same policy but many big US corporations allowed top bosses to continue into their 70s: in 2005, Exxon was led by Lee Raymond, aged 67. Knowing that BP was Browne's life, Sutherland had a suspicion that the CEO might not be eager to follow company tradition when the time came. He discussed the matter with Browne but while the CEO indicated he did not expect to stay on, there appeared to be a lack of conviction in his reassurances.

As it happened, Sutherland's suspicions were well founded. Browne made it clear to friends that he had no intention of leaving if he could at all avoid it.[48] He enjoyed hopping around the globe in presidential style, entertaining and being entertained by heads of state, and wasn't sure what he would do next if he left. But he also knew Sutherland would block any attempt by him to stay on. There was one possible solution: what if Sutherland were no longer around?

In June 2005, Sutherland spotted an editorial in *The Times*'s business section suggesting that BP might consider the services of Tony Blair as its next chairman. Blair had just won a third term as prime minister and had said that he would not fight a fourth election. Pundits expected him to stay a couple of years at most. When he saw it, Sutherland immediately suspected the story had been initiated by someone close to Browne. 'This is the opening blow in the war,' he told a friend.

At a meeting of non-executive directors in Venice, Sutherland asked everyone for their views on Browne's retirement. Unanimously they agreed he should not be allowed to extend his tenure beyond 60. Some even thought he should go earlier, but his popularity with investors and the general difficulty of getting rid of a CEO made this an unrealistic option.

Sutherland prepared himself for a battle. After *The Times*'s commentary, he decided it was a conflict that would be largely fought out in the press. In an unusual move for a chairman, Sutherland retained his own press adviser, independent of the company press office. It was especially strange – and a mark of how much distrust pervaded BP HQ – considering that BP's press chief Roddy Kennedy was a fellow Irishman and a close friend of Sutherland's. Sutherland feared Browne might seek to use Kennedy's team to advance his own retirement-delaying agenda, something that would be a technical abuse of its role.

After the 'Blair for chairman' idea failed to gain traction, Browne began to discuss with his various internal and external advisers the idea that he himself might become chairman of BP when he stepped down as CEO in 2008. Of course, he wasn't terribly keen on being an invisible chairman and told friends he thought an executive chairmanship would suit him better.[49] On both counts, this would have represented a breach of UK corporate governance rules, which frown upon CEO step-ups and executive chairmanships. He was advised that this was unlikely to pass muster with the board.

The discovery of the true extent of the management failings behind Texas City put further pressure on Browne's position. He saw that if he wanted to stay on, he would have to become more daring. For years he had harboured a secret desire to build the biggest oil company in the world by merging BP and Shell. Undaunted by the opposition that some at BP had to such a tie-up, he decided now might be the time to put the plan into action.

In early 2004, Shell had shocked its investors by declaring it had overstated its oil reserves by around a third. The news led to the sacking of top management and a collapse in the company's share price. It also led to the elevation of Jeroen van der Veer to the top of the company. The Dutchman was about as much of a contrast to Browne as one could imagine: tall and gangly with large ears, a side parting that occasionally drooped down towards his eyes, and slightly awkward movements. He had an understated, at times almost apologetic, manner, which he said stemmed from his Calvinist roots. In his low profile and lack of colour or panache he was the quintessential Shell CEO.

In June 2004, Browne and van der Veer travelled to Lake Como for a meeting of the Bilderberg group, an informal collection of senior international business, military and political leaders that come together annually to share ideas in an off-the-record environment – or, if you believe the conspiracy theorists, to run the world. In a break from the meetings, Browne and van der Veer went for a walk by the lake. Browne struck up a conversation about the future of the oil industry. He spoke in cryptic terms. Nonetheless, van der Veer got the picture: Browne was interested in bringing the two companies together. Van der Veer entertained the conversation but made no comment that could be construed as supportive. Neither man actually uttered words such as 'merger' or 'tie-up' in reference to the two companies. Despite this, Browne, according to the account in his own memoirs, left the conference convinced that the Dutchman had agreed to 'deeply consider' a merger.

Over the coming year and a half, Browne would make further indirect advances to Shell and interpreted the absence of an outright repudiation of this contact as a sign that van der Veer remained receptive to a deal.

A merger would be the realisation of all Browne's ambitions. Combined, the two companies would easily surpass Exxon in market value and only Saudi Arabia's state oil company would be a bigger oil and gas producer. Browne calculated that $9 billion in savings per year could be squeezed out of the combined group. And the deal would also solve his retirement problem. Van der Veer was close to Shell's retirement age of 60 and had always been a reluctant CEO. He made no secret of the fact that he considered the high point of his career to be managing the Pernis refinery near Rotterdam, which he would cycle around on a bicycle. He would happily have stayed there, except that senior management told him his duty to Shell required him to accept a promotion, which subsequently led to the top job. Given all this, Browne could be pretty sure that he would land the top job in any merger. Best of all, the integration of the two companies would be so complex that he calculated it would require five years to bed down, easily taking him past Sutherland's own retirement.

But Browne had badly misread the Shell CEO. A tie-up with BP was the furthest thing from van der Veer's mind. Indeed, he was a little offended by Browne's opportunistic advance. 'You should never take advantage of another person's misery,' he told colleagues back in The Hague. Shell would remain in a weakened position for years following the reserves scandal but van der Veer felt the company was a fundamentally healthy one. Doing a deal at this time would mean asking his investors to sell out cheaply. In any case, the whole concept of a merger between the two companies was 'nonsense', he told aides. He wasn't alone in this view: Shell's directors were well aware a merger with Shell had been at the top of Browne's wish list since the acquisition of Arco. 'We always knew that it was in the back of his mind,' said a former British Shell director, 'but it wasn't in the back of ours.' In contrast to BP's optimism on merger synergies, Shell's executives did not believe material cost savings could be made by putting the companies together. Indeed, Shell suspected a tie-up could make future growth harder, since resource-holding governments might be reluctant to do business with such a behemoth. Given that countries like Saudi Arabia and Iran didn't like BP, Shell might even have a hard time holding on to its existing projects.

But a bigger problem than all these was that a merger with BP would mean the death of Shell's Dutch heritage. The company was the Netherlands' one true global corporate giant. Ever since its founding in 1907, the Shell group's British directors and shareholders, which controlled 40 per cent of the group, had fought to gain the upper hand over the Dutch side, which held 60 per cent. Merging with BP would mean the Dutch side was diluted to the point of losing its majority. Everyone in The Hague knew what had happened to Amoco's tall, shiny HQ in Chicago. They figured it would only be a matter of time before Browne shuttered Shell's modest gable-fronted HQ just a short distance from the Royal Palace. As van der Veer well knew, Shell's Dutch directors would never agree to such a deal and neither would the Dutch government.

Browne had previously said he was ignorant of the problems building up within BP's trading, refining and pipeline divisions because people were reluctant to pass on bad news. The Shell experience suggests that in his later years, Browne had simply grown deaf to signals he didn't want to hear. For over a year after the Bilderberg gathering, Browne seems to have remained ignorant of Shell's implacable opposition to the notion of a merger. In September 2005, he convened a meeting of the heads of BP's operating divisions at a hotel near his apartment in Venice. One by one, Manzoni, Hayward and the others were asked what they thought of the proposed Shell deal. In a perfect display of how all challenge had been dismissed from the executive board, each of BP's 'princes' stood up and supported the idea. Some expressed concerns that a deal may not be possible: everyone knew that anti-trust regulators in Washington and Brussels might seek to block it. But Browne reassured them that he could avert this by selling all of BP's refineries and forecourts in the US and Europe. The prospect of ditching 70 per cent of BP's employees didn't appear to faze him; tellingly, he had already decided that if it came down to a choice between keeping his own refineries or owning Shell's, he would choose to have Shell's.

Tony Hayward was the coolest of the team to the proposal.[50] Perhaps he realised that his division would face the burden of delivering most of the savings and would face the biggest operational problems. Nonetheless, Hayward must have known better than to oppose his boss. He gave his assent.

With his troops prepared, Browne got ready for a meeting of the full board, which was due later that month in Williamsburg, Virginia.

Meanwhile, Sutherland knew of Browne's plan, and was also rallying his own troops. The chairman did not share the CEO's confidence that a deal with Shell would be a financial success, nor that it would win regulatory approval. Quietly, he approached the non-executives and shared his concerns. He insisted that EU competition rules would not allow such a merger: 'I should know, I wrote them,' he said in reference to his stint as EU Competition Commissioner.

When the board meeting came, Browne put forward his proposal that management enter merger talks with Shell. One by one, his executive team were asked their views and, one by one, they professed support. Then Sutherland invited the non-executives to speak. With the exception of one, they all said they opposed the plan, judging it too commercially risky, doomed to failure and likely to waste vast amounts of management time.

Browne was furious. He had been totally outflanked by Sutherland. 'John couldn't match Peter's skills in backroom politics,' one director who was present said. The meeting was a fundamental blow to Browne, not least because it marked a dramatic shift in the balance of power.

Browne moved on to another attempted tie-up, this time a deal to buy a major stake in Chinese oil company Sinopec.[51] It would have given BP a major foothold in a fast-growing market, but yet again it looked as if Browne's deft touch for deals had left him. He misread what the Chinese side wanted – to gain access to BP technology without ceding market share – and was rebuffed.

As time passed, colleagues detected an air of possible paranoia creeping in. Browne told aides he thought Sutherland's opposition to him might reflect a latent homophobia. As a devout Catholic and a native of a country that had only decriminalised homosexuality in 1993 following pressure from the European Union, Sutherland was, at one level, an easy target for such an accusation. But his previous very visible support for Browne suggested the accusation said more about Browne's state of mind than Sutherland's attitude to homosexuality.

Browne's suspicion then fell on press chief Roddy Kennedy, who he felt was not being as supportive in delaying his departure as he ought to be. The two had long been close friends, but friendship would not get in the way of Browne's goal. Kennedy had to go. He ordered his new head of human resources, Sally Bott, to draw up a severance package, but the plan was never implemented. The last thing Sutherland was going to allow was to have his friend replaced by a drum-beater for Browne.

Thwarted, Browne hired an external PR agency, Finsbury, a unit of advertising group WPP, which had also devised the Beyond Petroleum campaign. He began pursuing increasingly desperate and transparent schemes to prolong his tenure. In early 2006, he gave a speech denouncing compulsory retirement ages.[52] He then reiterated the point in a series of newspaper interviews.[53] Meanwhile, newspaper journalists began receiving telephone calls that Sutherland believed came from Martin Sorrell, chief executive of WPP, and Roland Rudd, the boss of Finsbury, urging them to write stories arguing for Browne to stay on.[54] When, in the middle of 2006, analysts at Merrill Lynch published a research note urging BP to change its retirement policy to allow Browne to stay on, Sutherland exploded with anger. He concluded that people close to Browne had briefed the analyst and summoned the CEO to his office late one Friday night. 'John, you've got to stop doing this,' he said. Browne denied any connection to the campaign being waged in his favour. According to Browne's aides, Sutherland began bellowing at Browne and demanded he publicly name a retirement date. Sutherland's friends disagree, describing it as a respectful but firm discussion, during which Sutherland merely told Browne to end what he saw as a campaign of media and analyst briefings.

The following day, Browne attended the wedding of his communications boss, Anji Hunter. He was seated in the church beside Andrew Neil, the former *Sunday Times* editor and Rupert Murdoch confidant, who, at the time, was publisher of a now-defunct Sunday business newspaper, with the unimaginative name *Sunday Business*. On Sunday morning, the front page of Neil's newspaper bore the headline: 'Boardroom split at BP over Browne's retirement date'. The newspaper quoted a 'city source, familiar with the situation' as saying: 'Sutherland is bullying John into retiring. The board has yet to make a decision over what it wants, and investors are keen for John to stay.' The general tone of the article made it clear that the source was sympathetic to Browne. But the source had painted a rather inaccurate picture: while there was indeed a split on the board, it wasn't an even one. John Browne was in a minority of one.

Sutherland was apoplectic when he saw the headline. He called Browne back into his office on the Monday morning and gave him an ultimatum: announce the following day that he planned to stand down in 2008, or Sutherland would call a board meeting to discuss Browne's position. 'I'll think about it,' Browne replied. A decades-long veteran of the toughest international trade talks, Sutherland wasn't one to be fobbed off when he had the upper hand in a negotiation. 'John, there's nothing to think about,' he said. 'If you walk out of this room, I'm calling a board meeting.'

It was the end of the road and Browne knew it. He agreed to put out a statement the following day, announcing that he would stand down in late 2008. Sutherland agreed to allow him a few extra months so that he could see in BP's centenary year.[55]

That same day, other newspapers also ran stories on the spat. Some even correctly divined the board's thinking. One might have predicted the market would be spooked by the sudden prospect of Browne's departure from BP, but when trading opened its shares rose 2 per cent. After Texas City, the Alaska pipeline leaks and *Thunder Horse*, the Sun King's crown didn't shine so brightly any more.

When the dust had settled, Sutherland called WPP's Sorrell, furious that one of BP's largest suppliers had engaged in what he saw as a putsch against the board. He got straight to the point. 'I understand', he said, 'you've been calling newspapers in relation to corporate governance matters here at BP. I'm just wondering if that was in a personal capacity or as the CEO of a company which provides BP with $300 million worth of marketing services each year?'

Sorrell was surprised by the call, but in truth he had half been expecting to feel Sutherland's wrath in some form or other. He had run WPP for 20 years and so knew that highly charged business battles, like military ones, were usually followed by score-settling. He had aligned himself closely with Browne and Browne had lost. There would obviously be consequences. Still, he tried to calm the waters. 'You've won,' he conceded, and suggested, 'Let's forget about it.'

But Sutherland wasn't in a forgetting or forgiving mood. In the years that followed, WPP's business with BP suffered, while Rudd, the head of Finsbury, was said to have been banned from even entering BP HQ again.

Ignoble end

In the end, Browne did not get to see BP's centenary.

After three years together, Browne and Chevalier broke up. Browne agreed to help the young Canadian's transition to singlehood by paying his living expenses for a year. But at the end of that year, Chevalier had failed to renew his vague familiarity with work. He continued to send Browne requests for money, and after a while these began to carry a hint of threat. Browne ignored the letters.

Then, in January 2007, while Browne was holidaying in Barbados, he received a call from Roddy Kennedy. Chevalier had sold his story to a newspaper and the paper sought comment on a number of allegations that represented the thinly veiled public interest justification for the story. The claims were that Browne had misappropriated BP resources for the benefit of his young lover by allowing Chevalier to fly on the company jet to Venice and by giving him a second-hand BP laptop.

Browne had given much thought to coming out in the past. In the end, he had always decided against it because he couldn't bear the snide remarks he felt would inevitably appear in the newspapers. This exposé was all Browne's nightmares come at once: the fear that his carefully honed reputation could be denigrated, the fear that his position at BP might become untenable, and even the fear of betrayal bred into him by his mother. He decided to seek legal advice and called Schillings, a libel law firm. Predictably, the lawyers advised him to seek an injunction blocking publication of the story.

The case was scheduled to be heard by Justice Eady, and Browne, supported by his friend Sorrell, felt confident of prevailing. Eady had a long record of awarding injunctions against newspapers. But Browne had lied in court documents, repeating the fib he had told friends for years about meeting Chevalier in Battersea Park. His case was thrown out and he was given a stinging rebuke. Browne appealed and lost again.

Browne told Sutherland about the kiss-and-tell story and his plea for an injunction, but didn't share other details, such as the lie, claiming that the judge had insisted the case be kept secret. As a former lawyer, Sutherland doubted that a judge would preclude the party seeking to keep a matter private from sharing information about the case, but he didn't press the matter. Even when Browne told him that all his efforts to secure an injunction had failed and that the story was about to break he did not mention the lie.

Thus, ignorant of the lie, Sutherland assured Browne there was no need to stand down. But the full truth was against the CEO. When Browne consulted Alan Parker, the head of Brunswick Group, London's largest financial PR agency, and told him the full story, he was advised that his position was untenable.

That afternoon, Browne walked out of BP's head office for the last time as CEO. It was a distasteful end. BP had by this stage investigated the claims of abuse of BP resources in the story, and Browne was asked to pay the cost of Chevalier's travel on the company jet, in line with company policy. The board agreed, however, that Browne's assistant having done some secretarial work for Chevalier and his use of the discarded laptop did not represent theft of company resources. These allegations were repeated in the story but received lower billing than details of Browne and Chevalier's dinner with Tony and Cherie Blair, their visit to Elton John's flat and other details of their jet-set life together.[56]

Of all the stories I ever wrote about BP for Reuters, Browne's departure was probably the most widely picked up by our media clients worldwide. Journals from Calgary to East Asia led on it. The company was at a low, but fortunately it had a new boss, already waiting in the wings: a new boss with grand plans.

4

The Perfect Candidate

A sense of anticipation filled the room on the sixth floor of BP's headquarters at St James's Square. Waiters in dark uniforms served hot coffee, warm pastries, fruit pieces on tiny bamboo spears and miniature bagels with smoked salmon and cream cheese. But the reporters present largely ignored the refreshments. They chattered, their eyes flickering incessantly towards the entrance.

The room on the top floor overlooked a small public park, St James's Square Gardens. Greenpeace protesters regularly climbed its trees to hang banners lambasting BP for its environmental record. But this was a frosty morning, and the branches were empty.

As the appointed time for the press conference grew near, Tony Hayward appeared, walking a few paces in front of his boss, for now, John Browne. China cups clattered into their saucers as reporters abruptly ended conversations with more lowly officials and angled towards the star attraction. It was 6 February 2007 and BP was unveiling its full-year results for the previous year. More importantly, the company was also unveiling the man who would succeed Browne as chief executive of BP that summer.

In his first public outing since the announcement, Hayward appeared very much the herald of a new generation. He had a fresh, boyish face, a full head of dark brown hair and a trim figure – he competed in triathlons – that belied his 49 years. The reserved Browne, by comparison, was showing the scars of his 12 years in office: his hair greying, his brow furrowed and, later, his need for glasses to read his statement. His grey pallor was all the more obvious next to Hayward's apparent tan. Browne was, as usual, wearing a tailored, Savile Row suit, handmade shoes and a sombre navy tie, which, as usual, was finished with a dimple under its four-in-hand knot. One got the sense that Browne might have found Hayward, with his off-the-peg Italian suit, loud pink tie (sans dimple), bright red cufflinks and chunky watch, a little gauche. But it didn't matter what Browne thought any more.

As he walked in, Hayward appeared to have a swagger about him. Maybe it was simply the way he wore his suit jacket unbuttoned; Browne was meticulous about keeping his closed. Either way, had Hayward felt a little pleased with himself that morning, it would have been entirely understandable: chief executive of Britain's largest company wasn't a bad achievement for a state-school-educated boy from Slough. The new role guaranteed Hayward enormous financial rewards, but also much more. A BP CEO was guaranteed an offer of a knighthood and was also a good bet for a peerage: Tony Blair had appointed Browne and his predecessor David Simon to the House of Lords. The top job had long been on Hayward's radar, and his appointment in 2003 as head of exploration and production, the company's main profit centre, had put him in pole position – and he knew it. Six months earlier in the same room, as the race to replace Browne was intensifying, I had asked him how the selection process was progressing. He scowled at me. 'I'm just focused on running E&P; it is 80 per cent of the company, you know,' he snapped before storming out of the room. The implication was clear to me at least: he was already doing most of the job, and the fact that he then had to compete for it hacked him off.

As it happened, the race for the top spot wasn't much of a competition. Manzoni had been ruled out by Texas City, while the other candidates were also-rans. The strongest among them were Bob Dudley, then head of BP's Russian joint venture, TNK-BP, and Iain Conn, a roving director who had formerly headed the chemicals division and who had impressed the board with what they considered his 'courageous' recommendation to offload that unit.[1] Dudley, one of the few former Amoco executives who had thrived at BP, was highly respected but hobbled by a low profile in the UK. His name came up a few times but, according to non-executive directors, he was never seriously considered. Conn, who had come up through BP's oil trading unit, was seen as having a good strategic mind but also as having tried a little too hard. 'He had manifest intelligence but also manifest ambition,' said one non-exec. Hayward's number two at exploration and production, Andy Inglis (pronounced 'Ingalls'), had also been shortlisted but was never really a serious contender.

This wasn't to say everyone inside and outside BP thought Hayward should get the job. His record at exploration and production had not been unblemished – the Alaska oil spills and *Thunder Horse* disaster had happened under his watch – but fortunately for Hayward the man running the selection process, Peter Sutherland, did not appear too bothered about the soiled tundra. In unscripted remarks at BP's 2006 AGM, Sutherland described the January 2006 spill, the worst ever on Alaska's North Slope, as 'extremely low', and understandable since 'sometimes corrosion is extremely difficult to establish'. Of course, Hayward could argue that the problems in Alaska, and with *Thunder Horse*, related to decisions that predated his appointment. In any case, the board took the view that Hayward's time in the upstream had been a success and elected him unanimously. 'He was a first-class explorationist and the company was an exploration company,' another board member bluntly remarked.

But Hayward had one weakness that no amount of spin or justification could brush over: he was not good at performing in public. He had competently represented BP's exploration in South America in the 1990s, but this was relatively minor stuff. As chief executive he would be expected to represent BP at the very highest level politically and publicly. He would face a scrutiny of intense proportion. Was he up to it?

Like many, I had my doubts. First, he appeared to have a tendency to speak out of turn. His comment about running most of the company was an example: it made him look prickly and petty. Even in prepared remarks, he appeared to struggle to get his message right. The previous December, shortly after he learnt he was to be the new CEO but before it was announced, Hayward posted a memo on BP's internal website, giving his thoughts on BP management practice. BP's management, he said, had made a 'virtue out of doing more for less', adding: 'The mantra of "more for less" says that we can get 100 per cent of the task completed with 90 per cent of the resources; which in some cases is okay and might work but it needs to be deployed with great judgement and wisdom . . . When it isn't, you run into trouble.'

The *Financial Times* got hold of the memo and published it under the headline 'BP memo criticises company leadership', and suggested the comments might be seen as a criticism of Browne, whose cost-cutting was, at the time, being blamed for causing the Texas City blast and the Alaskan oil spills. It's hard to see how anyone could read it any other way, but Hayward wrote a letter to the newspaper denying the comments were any such thing. It was the start of a long line of complaints about what he considered media misinterpretations of his remarks.

Journalists, analysts, investors and employees all scrutinise chief executives' comments for meaning, because a message may mean one thing to one stakeholder and another to a different one. And in some ways, it seemed to us that Hayward lacked the gravitas expected of someone who might soon control the livelihoods of 100,000 employees and impact hundreds of thousands of others who bought the company's product, lived near its facilities, relied on it for dividends and extended it credit.

Another example that sticks in my mind happened on my first visit to the sixth floor of St James's. It was at one of the lunches Browne hosted every six months for about a dozen selected journalists and his top executives. The meals were gourmet four-course affairs, served with fine wines, on inlaid tables with heavy silver cutlery and lead crystal. At one point, slightly uneased by the stiff occasion, I clumsily clipped my enormous wine glass against the side of a plate, prompting a loud and elongated ringing to emanate across the room. Everyone present respectfully ignored the beautifully pitched noise and continued to listen to Browne's pious meanderings on the state of the oil industry. Everyone except Hayward. As I tried to silence the glass with my hand, I heard a sniggering coming from down the table, and I turned to see the head of E&P grinning like a schoolboy. At the time I found Hayward's reaction amusing and even endearing, but looking back it made me think he might lack seriousness and self-control.

Part of the problem with Hayward was that he remained shy. This shyness sometimes made him appear arrogant or uncaring when he was simply uncomfortable. He didn't seem comfortable standing up and representing BP to the outside world. He certainly didn't seem comfortable representing himself to the outside world. Would he really be able to project the message BP wanted to get across? That these were key elements of the job didn't seem to bother the directors. After 12 years of Browne, board members felt they had seen enough of the celebrity CEO approach. 'We weren't looking for the media image type, we were looking for the practical implementation type,' said one director.

Hayward promised the directors he would focus on the brass tacks of the business rather than being an international statesman, seeking to wow people with transformational mergers or pursuing high-profile but unprofitable green energy ventures. 'BP makes its money by someone, somewhere, every day putting on boots, coveralls, a hard hat and glasses, and going out and turning valves,' he said a couple of years later.[2] 'That's how we make our money. And we'd sort of lost track of that.' The board bought his pitch and so did the investors. On the day his appointment was announced, BP's shares rose 5 per cent.[3] Financial analysts almost unanimously welcomed the appointment of a safe pair of hands.[4]

At the results presentation in February 2007, after Browne had done his bit, Hayward stood up at the podium and outlined the agenda for his tenure. He could not have been plainer. 'My priority is simple and clear,' he said. 'It is to implement our strategy, by focusing like a laser on safe and reliable operations.'

Everyone present nodded. It seemed a reasonable plan. The question was: how to achieve it?

The new old plan

Browne's premature departure following Chevalier's kiss-and-tell story pushed Hayward into the top job ahead of schedule, on 1 May 2007. Head of press Roddy Kennedy dispensed with the usual corporate tradition of taking a new CEO to do a round of interviews with the business press, CNBC and other financial media. The excuse was that Hayward's quicker-than-expected promotion meant he needed time to get his feet under the desk. Reporters on the oil beat suspected it also reflected Hayward's desire to keep a much lower profile than his predecessor.

Around St James's Square, other signs of a change in style were also evident. Hayward kept his old office on the fifth floor of the building. He converted Browne's suite of offices – known internally as 'the West Wing' – into meeting rooms. He removed the decorative 'eco' photographs hanging around the building, depicting objects such as leaves with droplets of water on them, and replaced them with blown-up photographs of oil rigs, pipelines and refineries. The security guards were told to bin the smart beige suits they had been issued and given more conventional dark replacements.

Hayward's unpretentious style appealed to many within BP, who had tired of Browne's imperial manner. 'He'd say "Call me John",' said one US employee, 'but you knew he really wanted to be "Lord Browne".' 'Tony' was altogether a different type of person: a beer drinker and a sports fan, and very much what the British call a 'bloke'. 'He probably felt more comfortable with blue-collar workers than he did with politicians and prime ministers,' said one executive director.

A fresh atmosphere pervaded the boardroom. Hayward ended Browne's practice of 'topping and tailing' his executive team's presentations to the board of directors, which some directors had considered an irritation. He also reduced the differential between the CEO's pay and that of his executive team. In 2006, Browne's salary and bonus had been two and a half times that of the head of E&P. In 2009, Hayward's salary and bonus was only 42 per cent above that of Andy Inglis, then head of E&P.[5]

Another area in which he was determined to set his tenure apart from Browne's was in its extent. He made it clear to all around him that he would stay no longer than five years as CEO. 'I'm only going to do the job for five years because more than five years, the power corrupts you and you fail to be effective,' he said.[6] It wasn't unusual for a newly appointed CEO to name his departure date. The logic was that one could not trust oneself, in such a job, to know when it was time to leave. In naming one's departure date at the outset, the risk of staying too long was avoided. Even so, a five-year term limit is unusually short.

The self-imposed deadline made Hayward keen to do things quickly and created an incentive for him to focus on short-term results. Certainly those around him felt his sense of urgency, exacerbated by the predicament in which BP found itself: its oil and gas production was falling and it was lagging behind peers on financial performance. Every Monday, when the top executives met and went through the latest figures, Hayward expressed his frustration. 'We can't go on like this,' he repeatedly told his top team.

Hayward did what many CEOs do in such a situation and called in the management consultants, in this case Bain & Company. The report Bain subsequently delivered described BP as too complex and bureaucratic. Browne's structure of decentralisation had led to duplication of efforts across the company. Each unit had replicated identical functions in design, operations and even media relations. Hayward believed this led to a massive overhead and slow decision-making. He also believed it contributed to BP's safety problems. Alaska and Texas City had occurred, in his view, not because of cost-cutting, but because people failed to appreciate the risks they were running in their businesses: quite simply, all the chatter between the multiple interfaces distracted people.[7] The question now was: what should Hayward do to improve day-to-day performance and ensure people knew how to recognise and better manage risks?

One possibility was to reduce the extent to which BP outsourced work to contractors. The concern was that, in relying so heavily on outsiders, BP had allowed its own technical skills to lag behind its rivals', and that this was contributing to the project delays and accidents. If BP brought more work back in-house, its technical skills would improve and projects would be both safer and built on time. But Hayward was cool on this idea. It would involve spending a lot of time and money, commodities he didn't feel he could afford to waste. When asked by an analyst at an investor presentation whether he would bring more previously outsourced work back in-house, Hayward replied, somewhat enigmatically, 'I think, not taking service providers in house, but what we have been doing is building greater capability to provide closer oversight of our supply chain, so greater capability at the front line to oversee the activity rather than bringing it back in house.'[8]

A more dramatic suggestion, mooted by an Alliance Bernstein oil analyst named Neil McMahon, came in the area of corporate structure. McMahon had started his career as a geoscientist with BP before moving into management consultancy, and at McKinsey had consulted for one of the firm's biggest clients: BP. He was one of the few people in the City who actually understood how BP worked. Shortly after Hayward was made CEO, McMahon penned a research note calling on BP to drop its 'hub and spoke' decentralised business model in favour of a centralised system, like Exxon's. 'We conclude', he wrote, 'that BP's de-centralized organizational structure, which had been fit for the purpose of cost cutting and for merger activities during the 1990's, is no longer fit for the purpose of growth implementation and the delivery of the current strategy today, and should be modified.'[9]

Industry executives believed Exxon managed to deliver projects on time and on budget, and without any of its refineries blowing up, because it had a clear functional chain of command made up of people who were experts in their particular field. All drilling, for example, was managed by a chief driller. This person had a deep technical understanding of drilling, honed through experience. Meanwhile, at BP, the head of each project would dictate his own drilling strategy. Browne's idea in building this structure of independent business units was that the project manager had the incentive to employ the most efficient means of building or maintaining a project. Why limit his choice to one internal expert, when a whole marketplace of possible suppliers and advisers was available? But the problem with Browne's idea was that it didn't work in practice. Managers were all commercial people, and they moved jobs every other year. They did not have sufficiently deep understanding of the systems and technology they were managing to know where costs could safely be cut and where they could not. McMahon figured switching to the Exxon functional model would boost efficiency, but everyone knew it would be an expensive and drawn-out process. Like the proposal on outsourcing, Hayward rejected the idea.

Instead, he decided to pursue something of a halfway house. He decided to reduce the independence of the business units and strengthen the core technological functions within BP. He would appoint powerful heads of drilling or technology, who would come up with a set of rules for how tasks should be completed. This ranged from procedures for building a production platform to ones on restarting a refinery after a shutdown. Having a defined set of practices would, Hayward believed, allow him to achieve a number of positive outcomes. First, disasters would be avoided because risk would be identified and managed better. Secondly, the cost of building projects would be reduced by standardising procedures. Under BP's decentralised structure, the company was forever employing a hundred different solutions to the same problem. Hayward figured that if a manager planning to bring a new field on line was forced to use the standard BP platform rather than try and invent his own, build costs would fall.

BP described its appointment of senior functional leaders and insistence on standardisation as taking a leaf from Exxon's book.[10] In truth it was, at best, Exxon-ish: a small tweak in an otherwise intact structure that continued to provide managers with incentives to focus on short-term results. 'Tony has said he would like to have more of the Exxon culture in BP,' said one senior safety executive, 'but I don't think he knows what it is.' Hayward wanted to echo Exxon's profitability but not its strict work practices. Indeed, the changes he instituted in many respects enhanced BP's fly-by-the-seat-of-your-pants culture. Since he was appointing strong new heads of functional areas to set policies on how to drill wells or build refineries, he could reduce the amount of people checking up daily on each other's work.

Finally, Hayward decided to ram these policies home with aggressive new remuneration packages that combined ambitious targets with highly variable bonuses. He figured that with these ideas, and without the distractions of Browne's flights of green fancy, BP would become an industry powerhouse again. In fact, he was simply making a bad situation worse.

Phoenix rises, again

Shortly after he was appointed, Hayward cancelled the investor roadshow that BP traditionally held in the second half of each year. Instead he set off on a tour of BP's offices worldwide to hold a series of 'town hall' meetings with staff, to sell his message.

The staff were not impressed with what they heard. Hayward's message of increased standardisation, cutting out layers of bureaucracy and simply pushing harder didn't really strike people as a strategy. They kept on asking him what his vision was for BP. 'We have to earn the right to have a vision,' was his rather uncertain reply at one town hall.

Hayward decided the lack of response to his message was due to the fact that BP employees didn't appreciate how bad a situation the company was in. He started to emphasise his message more strongly. At a town hall in Houston he told staff that BP's performance was 'dreadful'. BP's shares dropped 3 per cent on news of the briefing and prompted renewed claims from BP that the CEO had once again been misinterpreted.[11]

Hayward was unabashed. He decided he needed a big event to ram his new message home. But where to hold it? If one had asked any long-serving senior BP upstream executive in 2008 to name the single most significant corporate event in the previous 18 years, they would, to a man, have named Browne's 1990 conference at the Phoenician Hotel, Phoenix. If Hayward wanted to make a statement, he could choose no more obvious a location for his bash.

The old hands who gathered at the Phoenician in March 2008 found a much enlarged facility than the one they had visited in 1990. Another 100 acres had been added to the grounds, allowing the construction of a cactus garden, floodlit tennis courts and a croquet lawn. Inside, the hotel had additional rooms but the basic structure was the same and the décor was a familiar mix of marble and gilded fittings. Not that there were many old hands around who remembered the 1990 meeting. Hayward's continual harping about how he had inherited a broken company left them feeling less than valued and most departed. 'It was his Oedipal period,' said one former Browne aide, referring to the Freudian theory of defying one's father. Indeed, Hayward and Bob Malone, president of BP America, were probably the only two executive board members who had been present at the 1990 event, Manzoni having left a few months earlier to become chief executive of Talisman Energy, the company that had been formed a decade and a half earlier by the spin-off of BP's Canadian assets.

Attendees – this time there were 500, from across all BP's divisions – flew in on Sunday night and stayed until Wednesday. They were treated to two days of presentations from top management that left no one in any doubt as to the bad shape BP was in, and what Hayward wanted them to do about it.

To set the scene, one of the first presenters wasn't even from BP. Hayward was normally disparaging of financial analysts, dismissing them once as 'people who like to fill in spreadsheets'.[12] One would have expected him to be even less enamoured of an analyst who had recently cut his price target on BP's stock and written that, 'Within a large cap European context, the more we hear from BP . . . the more attractive the investment case appears at [French oil company] Total.'[13] But, as it happened, a BP bear like Neil Perry, from investment bank Morgan Stanley, was just what Hayward needed. Perry stood up and said that BP was failing at finding oil, failing in refining, and predicted that, unless the company changed dramatically, 'BP will not exist in four to five years' time in its current form.'

The message was that BP was at risk of either going broke or, more likely, being taken over. (Hayward would later say[14] that his own analysis on becoming CEO was that he had a timeframe of just two years to turn things around.) Hayward told the assembled managers that BP had become 'a serial underperformer' that had 'promised a lot but not delivered very much'. There was no time or place for people who were not with the programme. 'Either get on the bus, or get out. The bus is rolling,' he warned.

BP needed to become more efficient. The overhead would be cut and layers of oversight would be culled. 'We have too many people doing checks on the checkers,' said Hayward. 'Assurance is killing us.' He wanted people to be brave and make decisions, rather than allow themselves to be held up by people looking over their shoulders.

The Phoenix meeting sent a chill through BP. Managers saw that the company was in a mess and that the plan for getting it out was little more than hard graft. One subsequent departee observed that, under Hayward, 'the culture became much more macho . . . There was much more testosterone around.' (Indeed, outside commentators began to note that a number of high-profile senior women left the company. Cynthia Warner, group vice president of health, safety and environment and technology for refining and marketing, left, as did Cinzia De Santis, director of safety culture and leadership, and Deborah Grubbe, vice president, group safety. Vivienne Cox, head of alternative energy and the most senior woman at BP, also left, as did marketing executives Anna Catalano and Iram Shah, as well as group vice president in the downstream [and first female Browne turtle] Linda Adamany, and the CEO of BP Angola Mary Shafer-Malicki. In a speech at Stanford, Hayward had hailed Patti Bellinger, BP's former director, culture and diversity, as an inspiration to him. She also walked.)

Hayward called in KPMG, with a remit that the consultancy would later describe to prospective clients as extracting 'private equity style cost cuts', although BP itself announced neither KPMG's appointment nor its remit.[15] Hayward said he was cutting back on the superfluous bureaucracy that had built up under Browne, and would redirect resources towards the front line. But BP already had the smallest head office of all its peers: a modestly sized, six-storey building that housed only a few hundred workers. It begged the question: what constituted overhead and what constituted front-line services?

Communications didn't make the cut as a front-line activity, even though the right to exploit natural resources hinged on the goodwill of the public and governments. Hayward axed his head of communications, the former Blair chief of staff Anji Hunter, and did not replace her. The Washington lobbying budget was cut and the liveried waiters, fresh fruit and finger food at press conferences were replaced with cheap biscuits and serve-yourself flasks of weak coffee. Renewable energy also failed to qualify. A plain-thinking upstream person, Hayward didn't seem to believe in renewable energy or the soft benefits that Browne and others thought BP received from being seen as a green leader. Alternative energy executives said that, 'when challenged', Hayward would accept the science of climate change – but he didn't believe it was BP's problem to solve. He dismissed Browne's green investments as 'philanthropy'.

Hayward decided to whip the alternative energy portfolio into shape. This meant focusing on the natural gas-related activities in the unit. 'If you're going to work in alternative energy, work in something that matters to the firm,' he told one senior executive in the solar unit. He tried to offload the solar and wind units in stock market flotations, but was forced to hold on to the businesses when markets tanked in 2008.[16] Not prepared to keep throwing what he saw as good money after bad, he slashed investment in the alternative energy division by around a third. The renewable energies were hardest hit. Even concern about negative publicity couldn't stop Hayward from shuttering BP's US solar plants.[17]

Though clearly a front-line operation, refining also found itself in Hayward's sights. As a dyed-in-the-wool upstream man, the CEO didn't like the way refining tied up lots of capital and offered low returns. Iain Conn had replaced Manzoni as head of the downstream unit, and he found himself under immense pressure to make deep cuts. Earlier in 2008, he had acknowledged that, if only BP had operated as well as its best rivals, it could have made another $5.5 billion in refining the previous year. He attributed a big chunk of this 'performance gap' to 'uncompetitive overhead and support costs',[18] and set a course to close the gap by the end of 2011. By the end of 2009, he had clawed back $3.5–4 billion through savage measures that included moving 50 per cent of senior managers out of their jobs.[19] But it still wasn't enough for Hayward. In early 2010, he set Conn the target of taking costs to 2004 levels over the following two to three years.[20] Industry consultant CERA, which compiled a closely watched index on downstream capital costs, estimated these costs had risen over 50 per cent from 2004 to the first quarter of 2010. Some colleagues doubted Conn could ever make Hayward happy and believed his days were numbered.

But nowhere was immune from the cost-cutting drive, not even Hayward's beloved upstream oil and gas division. Andy Inglis had taken over as head of this unit after Hayward's appointment as CEO. A Northerner whose accent was still strong despite nearly 20 years in America, Inglis was feared for his fierce grilling; he was not someone people wanted to share bad news with. He saw himself as a passionate engineer. While convincing on one level – he had joined BP after receiving an engineering degree from Cambridge and, like his father before him, he had been made a fellow of both the Institute of Mechanical Engineers and the Royal Academy of Engineering – those who worked with him also said he was equally focused on the commercial side of the business. 'Andy always took close notice of the numbers,' said one business unit leader. He meticulously tracked BP's performance against rivals and used industry benchmarks as sticks with which to beat his subordinates.

Inglis rose to Hayward's challenge of cutting costs. BP spent $7–8 billion each year drilling wells and he told Hayward he could shave $500 million a year off the budget.[21] But how?

Almost all of this work was done by outside contractors so BP couldn't save much by cutting back on its own functions. Since contractors' rates were largely set by the market, effective haggling could only achieve so much. This left only one variable in which Inglis could make savings: time. It cost hundreds of thousands of dollars a day to rent a drilling rig, and other operating expenses could push the total bill to over $1 million each day. If the rig could be made to drill faster, the total cost would fall. There was even an industry metric, 'the number of days per 10,000 feet drilled', which could be used to measure efficiency in this area. Savings could also be made if BP cut the number of days on which a rig was not drilling – for example, when it was sitting idle waiting for a piece of equipment to be shipped from shore. Such 'non-productive time' (NPT) was measured closely across the industry. In the offshore sector, BP calculated industry NPT was running at 35 per cent in 2007, presenting a clear opportunity for savings.[22]

As oil prices fell in 2008, other companies also tried to cut costs, but BP claimed it had been 'more aggressive'. In February 2010, Hayward announced that BP's annual costs had also come down by $4 billion, helped by a reduction of around 10 per cent in head count (on top of job losses due to the sale of businesses). Management predicted even more cuts to come. 'It's a big machine and it can get fitter,' Inglis told me.

Safety first

Such extensive cost-cutting from a company with BP's track record might have been expected to raise alarm bells. But BP had an answer to such worries: OMS. The operational management system was designed to allow BP to attain the holy grail of the oil industry: good safety plus low costs. Hayward's heads of functions were told to provide 'enduring descriptions of "what good looks like"',[23] rules enshrined within OMS which would then guide workers in their jobs. Hayward reasoned this would make BP as safe and efficient as the best in the industry.

One of the problems with this strategy was that, when it came to filling key roles, Hayward tended to continue BP's practice of preferring more commercially minded people rather than technically experienced ones. For example, he hired Barbara Yilmaz as global head of drilling. Yilmaz had strong oil industry credentials: she was a qualified geologist and had a distinguished career running business units for BP. Her son was a petroleum engineer and she was married to a Turkish-American geo-engineer who was a 'world-renowned expert in Cone Pentrometer Testing and geotechnical drilling', according to Louisiana State University, which inducted him into its 'Hall of Distinction'. But when it came to drilling, Yilmaz was a relative novice. Little surprise, then, that when she discussed good drilling performance, she spoke of efficiency measures such as NPT. She told contractors at one conference that, unless NPT was cut, 'our projects won't be as robust. Over time it will affect how many projects we have going forward.' Some colleagues felt she lacked the technical skills to devise a comprehensive set of policies that could ensure safe drilling practices.

But even if people like Yilmaz were to devise good policies, the fact was that there was no way of making sure these were implemented. Hayward had already declared war on BP's internal audit functions, which he believed were expensive and slowed down decision-making. And when, in 2008, BP America president Bob Malone announced his retirement, Hayward cut a layer of oversight implemented by John Browne after Texas City and the Alaskan oil spills. To reassure the US authorities that BP was serious about health and safety, Malone had been granted the right to halt any work he thought was unsafe. A 'swat team' of former safety staff from Du Pont – thought to have the best safety standards of any company in hazardous industries – went around BP facilities checking that staff were following best procedures. All this was eliminated when Malone left; the job also changed, and his replacement, Lamar McKay, ceased to have the same level of operational authority as Malone, becoming more of a figurehead.

Similarly, when John Mogford, BP's global head of safety and operations, moved out of his role in late 2007, Hayward downgraded the position. Mogford had reported directly to the CEO, but his replacement, Mark Bly, did not sit on the executive board.

There was, however, one person who thrived under Hayward's leadership: human resources director Sally Bott. In her, Hayward found an eager ally to steer his performance-driven culture forward. Bott, an American, had joined from HSBC, and employees felt she had brought the hard, short-term results-focused investment banking approach with her. She wholeheartedly supported Hayward's head-count reduction drive, both within her own department and by orchestrating the mass sackings that Hayward's policies required elsewhere.

But her most controversial contribution was in helping devise managers' remuneration packages. Investors did not know it, but underpinning the billions of dollars in savings Hayward achieved was a system of bonuses that incentivised BP's managers to work quickly and cheaply and even, in some cases, with reckless abandon. Stretch targets were set and more variable bonuses than before were tied to them. As one senior executive described it, if the available data said the most one could possibly produce from the existing fields and planned start-ups was 3.9 million barrels, the target against which bonuses were set might be 4.2 million barrels. There would still be bonuses if one hit 3.9 million, or potentially even if one fell short, but it was now possible to receive dramatically higher bonuses than before by hitting the stretch targets. Thus began a continuous effort to go beyond what BP's own engineers considered physically possible. For senior drilling personnel, annual bonuses could add over $100,000 to salaries of around $200,000. Such bonuses were linked to how fast you drilled, measured in 'days per 10,000 feet of drilling'.[24] Because drilling too fast was dangerous, other companies were not usually in the practice of focusing on such a metric when assessing workers. 'We stumbled into incentivising people the wrong way,' admitted one senior executive who worked closely with Hayward.

There was a component for safety in the performance contracts managers agreed with their superiors – indeed, some managers would even find safety at the top of their list of objectives – but the safety element typically accounted for only 10 per cent of the bonus, and measured the wrong thing, according to the Presidential Commission investigation into the 2010 oil spill.[25] Drilling managers in the Gulf of Mexico, for instance, were gauged on the number of recordable injuries suffered under their watch. This was a personal safety metric of the kind BP had been criticised for relying on at Grangemouth in 2000 and at Texas City in 2005. Of course, staff were also told they had to comply with BP's drilling policies, which were intended to minimise the more serious process safety risks. But even if these had been devised by technical specialists to ensure they matched best industry practice, there was no one to check that the policies were actually being adhered to. Consequently, non-compliance with written policies was common.

The lack of focus on process safety led to a series of disagreements between BP's senior vice president for drilling operations for the Gulf of Mexico, Kevin Lacy, and managers like Yilmaz and Inglis. Lacy was a recognised industry authority in health and safety[26] and had been hired from Chevron, the second-biggest player in the Gulf, to improve and standardise BP's drilling practices. But he became convinced that BP 'was not adequately committed to improving its safety protocols in offshore drilling to the level of its industry peers', according to court documents filed by BP investors in early 2011, and about which BP declined to comment. Investors were ignorant of the problems in E&P because BP did not as a matter of course disclose details of its managers' performance contracts or news of Lacy's departure and the reasons behind it.

Indeed, it was unclear how much even Hayward knew about these matters since Inglis ran his unit with minimal oversight. Inglis believed Hayward should hold him to account for meeting the goals in his performance contract, but that the CEO should otherwise stay out of

E&P. When Inglis began lobbying Hayward to allow him to move the global headquarters of E&P to Houston, top managers questioned his claim that it would be cheaper to run the unit from Texas, wondering instead whether Inglis simply wanted more independence. The E&P move to Texas was eventually agreed, even though it put 5,000 miles between the CEO and the unit which, as he had once so forcefully reminded me, was responsible for 80 per cent of group earnings.

Hayward was right when he complained that he had received a broken inheritance, but his changes failed to address the fundamental flaw at BP: a structure that encouraged managers to put short-term financial goals ahead of the long-term health of the business and its employees. Employees still had incentives to take on ever more risk for the sake of short-term performance.

At BP, risk had no cost – at least not for top managers. Consequently they could in effect ignore it. By 2010, every time BP drilled a well it was like a trip to the casino with other people's money – a point made at the 2011 AGM in a message read out from Keith Jones, whose son Gordon Jones died on *Deepwater Horizon*. It read: 'You rolled the dice with my son's life, and you lost.'

Investors' favourite

In 2007, BP's profits dropped, despite a soaring oil price. Oil and gas production fell, for the second year in a row, and Hayward said that in future production would grow more slowly than earlier indicated. The news had a predictable impact on investors, and the company's shares lagged behind those of its rivals. Hayward received no payout under BP's long-term incentive programme for 2007.

But it wasn't long before the CEO's hard driving of staff started to turn things around. In 2008, production began to rise and profits soared 50 per cent, partly helped by rising crude prices. By comparison, Shell saw production fall and only a 14 per cent rise in profits. In 2009, BP would accelerate its operational improvement with an over 4 per cent rise in production. In the same year, Exxon reported flat output and Shell's production fell. BP told analysts it was enjoying a 'flywheel effect',[27] which would power earnings for years to come. 'We've had trouble forecasting ourselves as the momentum has picked up so much,' Hayward said in an interview.[28] The market loved the mix of strong production, good profits and big cost cuts. Morgan Stanley told clients that 'reputational and operational reliability issues are being fixed with a steely determination, in our view.'[29]

Hayward said he was changing corporate culture, something that anyone with the faintest understanding of large corporations knew took years to achieve. Nonetheless, barely 18 months after he arrived, the financial community decided it was 'mission accomplished'. Why were people so ready to accept such a simple explanation for BP's sudden turnaround?

Part of it can be explained by the stock market's inherent optimism. If it were not for the assumption that shares will go up over time, there would be no stock market. Part of it is related to the fact that, with BP accounting for around 10 per cent of the UK's main stock index, analysts are discouraged from being too hard on the company. It would be a challenge to keep a job as an oil analyst in the City of London if BP's investor relations department stopped taking your calls.

Finally, the lure of the turnaround story also encouraged converts to BP's cause. Until the end of the twentieth century, research analysts had existed to help their colleagues in mergers and acquisitions (M&A) and equity capital markets (ECM) win mandates to advise on takeovers or stock market flotations. Investment banks offered companies the promise of positive analysis in return for landing the deal. This meant the banks got fees, the company got positive analyst coverage and therefore an increase in share prices, and management got big bonuses. It was win–win for everyone – except the investor who was left holding the stock when the market found out the company wasn't as wonderful as the analysts had said it was.

Following a shake-up in equity research regulations, analysts could no longer be paid by their colleagues in M&A and ECM, so they had to find a new paymaster. There really was only one other place to go: the investment banks' trading floors. By the time Hayward took charge of BP, analysts' role was to produce research notes which their banks' brokerage arms could send to their clients, the fund managers. Brokerages made money from fees charged to fund managers who traded through them. If the clients liked the research, they would trade through the bank. This model created less incentive to produce excessively rosy research, but it did create an incentive to focus on 'stories'. It was an analyst's job to generate reasons for clients to trade, and the best kind of story was a turnaround: clients needed to fill their boots and get on the train before it left the station.

The influence of such research notes extends far beyond the clients who read them, since they are the basis of most commentary in the business press. These newspaper columns are in turn the sources of the views expressed to an even larger audience by the red-braces-wearing commentators on consumer television channels such as the BBC.

The straight facts of the Hayward era tell a tale of harsh cost-cutting and woolly or even contradictory statements on safety. Nonetheless, the accepted wisdom was that Hayward had made BP both safer and more efficient. No wonder, then, that few investors begrudged him the near doubling of his pay between 2007 and 2009, when he netted $6 million.[30]

What lurks beneath

A general lack of transparency in the corporate world makes it hard ever to know what is going on inside a company. All information is controlled by the management, whose primary goal is to manage perceptions of their own performance. Obviously their primary duty is to protect shareholders' interests, but anyone who thinks this duty will ever trump an executive's desire to manage his or her reputation is naive. Indeed, information that might be highly beneficial to shareholders yet damaging to management is often withheld from investors on the grounds that releasing it would hurt shareholders. The term 'commercially sensitive' can be stretched to audacious extremes.

An example occurred in late 2008, in Azerbaijan. Early one morning, workers on a BP platform 100 kilometres east of Baku in the Caspian Sea were awoken by an alarm. Sensors had detected high levels of gas, and when workers looked over the side of the rig, they noticed bubbles emerging from the sea around them. It was a potentially catastrophic scenario. All 200 workers were evacuated and production at the platform, and at another one nearby, was shut down immediately. The BP manager responsible later said the company was lucky to get everyone out alive.

The facility was part of the million-barrel-per-day Azeri-Chirag Gunashli group of fields, one of the jewels in BP's portfolio and the main revenue generator for the Azeri government. In the coming weeks, BP sent down miniature submarines to investigate the situation on the sea floor. They found gas leaking up from around the well head, the result of a bad cement job. BP repaired the damage, at great expense, and production was finally restarted. Though it would have been of great interest to investors to know about shortcomings in BP's cementing abilities, the company didn't publicise the true cause of the leak. This remained a 'commercial secret' and so investors' faith in the company remained intact.

That said, the truth was not so much of a secret that the company couldn't share it with local US diplomats, something Western oil companies regularly do in order to keep diplomats sweet in case of disputes with the local government. The world only learnt the truth of the Azerbaijan gas leak after WikiLeaks published classified US diplomatic cables referring to the matter in late 2010.

There were also visible signs that Hayward's claims on efficiency and safety were not quite accurate. In October 2009, the US Occupational Safety and Health Administration (OSHA) hit BP with an $87.4 million fine for failing to address safety violations at the Texas City refinery. In the aftermath of the fatal explosion, OSHA and BP had agreed a programme of work to make the refinery safe. OSHA was now claiming that BP had not fulfilled its obligations and that these failings 'could lead to another catastrophe'. Dean McDaniel, OSHA regional administrator, said BP had started out well but that, about three years after the blast – around the time of Hayward's Phoenix meeting – things had started to wane. In the four years after the explosion, workers continued to die at the facility at the one-a-year rate that had shocked BP in the aftermath of the blast. The fine was the highest ever levied by the safety regulator, surpassing the previous record, a $21.3 million fine it levied on BP in 2005 for the Texas City blast itself. For a corporation with billions of dollars of profits each quarter, the fine was not the problem: it was the reputational hit that was potentially more damaging. A comment from Jordan Barab, Assistant Secretary at the Department of Labor, that 'there are some serious systemic safety problems within the corporation' was a direct attack on Hayward's claim to have fundamentally changed BP's safety culture.[31]

BP challenged the fine and the claim that it had not complied with the agreement. Privately, the company told anyone who would listen that the fine was a political move: Barack Obama had moved into the White House earlier in the year, and the new administration had been signalling to industry leaders that what the Democrats saw as lax enforcement of labour laws under President Bush was going to end. BP said that Obama wanted to send a hard message to industry and was using BP to do it, in the knowledge that BP could not strike back because of its record of accidents, and because it was a foreign company. 'We've become the whipping boy,' one director complained. The argument found favour in the UK, where scepticism about the US political and regulatory system abounds. The headline in the left-of-centre *Independent* declared that a 'Heavy whiff of politics hangs over BP fine', and noted that 'Regulators, whose senior staff are all political appointees, are acutely sensitive to these political winds.'

The fact was that, even if one stripped out the 709 citations OSHA announced in October 2009, BP's US refineries would still have been way ahead of their rivals in breaching regulations. Between June 2007 and February 2010, BP incurred 862 citations from OSHA, compared to 127 for Sunoco, the next worst offender, and 119 for ConocoPhillips, the third worst. Among themselves, some BP executives acknowledged that OSHA might be onto something. The regulator's observation that BP had slowed the rate of upgrade at Texas City coincided with Hayward ratcheting up the pressure on Conn to cut costs in the refining division, and the abolition of the internal audit function that had existed within BP America under Bob Malone. Whether there was a direct causal link will never be known as the case was settled.

Just one month after the record-breaking fine, investors received another sign that all was not well beneath BP's bonnet: the company reported another spill in Alaska. A two-foot-long crack in a flow line, caused by a build-up of ice within the pipe, spilt 46,000 gallons of an oil and water mixture onto snowy tundra at Prudhoe Bay. The reactions of the public and of investors were muted; BP's shares were unmoved. One analyst summed up the mood of the market: 'We would be surprised and disappointed if this proves anything other than an isolated incident.'[32] As it happened, it wasn't. A week later, another pipeline ruptured. Fortunately the spill was contained in an outdoor gravel area. Later in December, BP found a third leak.[33]

But the discovery of three leaks in a month failed to capture the public's attention. The press coverage of the leaks was sparse. News on the spills was available, as newswires like Reuters and Dow Jones covered it extensively. Their clients in the oil markets needed to know if there was a risk of Prudhoe Bay being shut down, in which case crude would spike as it had done during the 2006 closure. But media clients were not so interested. Perhaps this was because it was winter and Prudhoe Bay was enjoying 24-hour darkness, thereby limiting the opportunity for TV pictures or photographs. Perhaps it was because Prudhoe Bay was just too far away for anyone to get to or care about. Or maybe it was simply because the story just sounded like an old one. Either way, the BP juggernaut continued to roll on and investors were happy to believe that, yet again, BP had achieved the alchemy of cutting costs while improving safety. BP clawed its way back up the industry league table, retaking the number three spot it had lost to Chevron. Then, in early 2010, it overtook Shell to become Europe's biggest oil company and the industry's number two.

After living for years under John Browne's shadow, Hayward began to shed his image as the master's apprentice. He had a new-found confidence, reflected in the way he started to give interviews, even on television. And best of all, the financial crisis meant bankers had taken on Big Oil's mantle as the popular villains of the corporate world. Hayward bumped into Goldman Sachs CEO Lloyd Blankfein and J. P. Morgan's Jamie Dimon in New York when the bankers were fresh from a Congressional grilling. He joked that it was good to see someone other than the oil companies being trashed on the front pages of the world's newspapers.[34] Pretty soon, Blankfein would be thanking Hayward.

5

Macondo

In 2010, the Gulf of Mexico was the greatest prize in the oil industry. Saudi Arabia, Venezuela and Russia had bigger reserves, but they denied access to Western oil majors like Exxon, Royal Dutch Shell, Chevron and BP. The rise in oil prices over the previous decade had prompted a wave of 'resource nationalism' in Latin America, Africa and Asia, which saw countries increasingly reserve their richest fields for development by their own state oil companies. On the rare occasion when a company was invited in, the terms were invariably bad and prone to change. Russia, for example, levied tax at a rate of 90 per cent on profits above $25 a barrel,[1] and had stopped selling foreign companies the rights to large fields. Venezuela had redrawn all the contracts it had signed with foreign oil companies to shrink the profits they could earn and to give the state oil company PDVSA a majority share in all projects.

The Gulf of Mexico, on the other hand, offered a unique blend of stable politics, open access, low taxes, an absence of labour unions and the potential for multibillion-barrel oil finds. The US government estimated total reserves at 45 billion barrels. These factors made Gulf of Mexico crude the most profitable in the world. In late 2007, when crude was trading for around $75 a barrel, Shell had revealed that it was earning $2–4 a barrel onshore in the Niger Delta and $10–12 in the North Sea, but that the margins in the Gulf were $20 a barrel.

The other great attraction of the Gulf was that, after the government had sold you a licence, it didn't try to micromanage your affairs. The US regulatory environment was one of the most lax in the developed world, far behind countries such as Britain and Norway, which are seen as having the most stringent regimes. Underpinning America's accommodating approach was the long-held view that offshore drilling was safe. In 1978, Senator J. Bennett Johnston, a Louisiana Democrat, declared that 'The so-called danger from oil spills has simply not been proved. Not only has it not been proved, it has been disproved.'[2] Over the following 31 years, the US Gulf of Mexico had no major oil spills. Even politicians who leaned more towards green energy accepted that drilling for oil was safe: 'Oil rigs today generally don't cause spills,' announced President Barack Obama in early April 2010.

This wasn't to say that the industry regulator, the Minerals Management Service (MMS), gave the industry everything it wanted. Indeed, oil men still saw room for improvement, in their favour. 'The regulatory approach has to be less precautionary and based more on the cost–benefit ratio,' Exxon chief executive Rex Tillerson was quoted as saying in an interview he gave to a rival oil company's in-house magazine on 19 April 2010 (and luckily for him, given what happened the following day, the comments were not reported in the mainstream media).[3] Fortunately for the industry, the MMS was a listening regulator. The regime relied largely on self-regulation, which is to say that the industry itself determined what constituted best practice. Guidelines were published by the industry lobby group but were not binding. This approach had facilitated the breakneck development of the Gulf of Mexico.

Pushing into deep water

The first time men drilled for oil in water was in the 1890s at a lake in Ohio. A few years later the first wells in the sea were drilled, from piers, near Santa Barbara, California. The first time oil men got their feet seriously wet was in the 1930s, when two independent explorers, Pure Oil and Superior Oil, joined together to buy Gulf of Mexico State Lease Number 1. The block lay a mile and a half from shore, in 14 feet of water, requiring the construction of a freestanding structure that could be described as the ancestor of today's offshore oil rigs. Over the coming decades, explorers pushed further and further offshore.

Crude oil is formed from the decomposition of organic matter under pressure, over millions of years. This source matter usually originates on land: when oil is found offshore it is because rivers brought the source matter there. Consequently, offshore oil needs a continental shelf and it needs rivers. The bigger the rivers and the bigger the land mass, the bigger the potential finds, and the further from shore they may be located. Given the size of the North American land mass and the breadth of the Mississippi river, oil men knew the limits for exploration offshore Louisiana were more likely to be technical than geological.

In the 1970s, the development of new seismic surveys and improvements in platform design allowed a push into deep water – depths of over 1,000 feet. Shell led the way. The only other company with the size and technical ability to handle the engineering challenges involved was Exxon, but the sums were so potentially ruinous that the conservative Texas-based company shied away. Even later, the company would remain a relatively small player in the Gulf; in so far as it did participate, it was often as a financial investor in wells rather than as an operator. BP, meanwhile, was nowhere to be seen in the early years of deepwater.

In 1988, amid low oil prices, Shell sought to mitigate the risk in its latest deepwater adventure, the Mars prospect. It sold a 28.5 per cent stake in the project to BP, giving the latter its first taste of the deepwater.

Shell subsequently struck oil – over 500 million barrels of it – providing a much-needed windfall for the then-troubled British oil giant. It also provided a useful schooling in deepwater technology for BP's engineers. Nonetheless, BP did not jump immediately and wholeheartedly into the Gulf. In the 1990s, John Browne was more interested in exploiting reserves being opened up by politics, most notably in the former Soviet Union. This drive led to big positions in Azerbaijan and Russia, but little activity in the Gulf.

The big uptick in BP's Gulf of Mexico portfolio came with the takeover of Amoco in 1998. Overnight, the company became one of the biggest players in the region. Meanwhile, as oil prices dropped in the mid-1990s, Shell had begun to shy away from the Gulf, finding it hard to justify the high costs in a world of $10-a-barrel crude. By the turn of the millennium, BP was seen as the leader in the Gulf, much to the chagrin of its rival. Adding insult to injury, BP kept poaching Shell's staff. The Anglo-Dutch company put all its drilling engineers through a rigorous two-year training programme, during which bright young engineering and science graduates were inculcated with Shell's technical values. They received both practical and classroom training and had to sit an exam at the end. If you didn't pass – and around 20 per cent didn't – you had to find a new job. BP did not have a comparable programme and, whenever possible, it hired the bright-eyed young things from Shell.

As BP made new finds in the Gulf, it aggressively added new licences on top of those inherited from Amoco. Shell's engineers grumbled that BP was growing too fast and questioned whether the company had the technical skills and infrastructure to support their push. 'BP say they can deploy technology as good as Shell's in half the time at half the cost,' moaned one. 'How can they do that? Only because they're prepared to tolerate a risk profile we wouldn't.' That said, Shell's own view of itself as a paragon of engineering excellence was being challenged at this time by its own reluctance to spend money. In the North Sea, workers nicknamed the company's failure to properly maintain platforms as a policy of TFA (touch fuck-all).[4]

In the early 2000s, the buzz in the Gulf was around the ultra-deepwater: depths of 5,000 feet and more. BP powered ahead on this new frontier, making new discoveries to which it gave such exotic names as *Na Kika*, *Mad Dog* and *Crazy Horse* (later renamed *Thunder Horse* following complaints from descendants of the dead Indian chief).[5] The strange names had a practical purpose: securing oil licences was a highly competitive affair and interested parties didn't like to show their hands publicly, so they ascribed code names to the exploration blocks they bid for. It was critical the code name had no geological or geographic relevance to the block being bid on, so the wackier the name, the better. Some companies, including BP, held auctions where employees could bid to name a prospect, with the proceeds going to charity.

The Amoco and Arco deals made BP the largest oil producer in the US by the turn of the millennium. By 2010, it was also the top producer in the Gulf and owned more deepwater and ultra-deepwater leases than any other company, suggesting it could maintain its lead for years to come. The importance of the Gulf to the company was clear from its weighting in BP's exploration forum, the 17-seat body headed by Andy Inglis, which managed the exploration and production unit: of the ten regional seats,[6] representatives of the Gulf occupied three. The most any country outside the US had was one. The forum channelled more cash to the Gulf than towards any other investment location.

Carefully polished image

BP was relying on the Gulf of Mexico to drive its growth in the second decade of the twenty-first century. It used the area as a calling card in other countries that wanted to open up offshore fields; even resource holders who generally favoured keeping big fields for their national oil companies were often forced to invite the big international oil companies in to help tap more complex resources, such as deepwater projects. The Gulf was successfully used to highlight to countries such as Russia, Angola and Egypt why they should do business with BP.

As part of its ongoing effort to showcase its Gulf of Mexico operations, BP invited a couple of dozen analysts for a two-day visit to Houston in May 2009. The first day was taken up with a series of presentations at BP's West Houston campus: a group of high-rise buildings set in landscaped gardens, with man-made ponds filled with ducks and geese and lawns neatly manicured by Hispanic gardeners. Each block had its own high-rise car park, and, as is the case in most of Houston, walkways connected the two, so that workers were shielded from the punishing Texas sun.

The presentations were led by Inglis, who was fast making a name for himself with the analyst community, although his manner could hardly be described as charismatic. Indeed, the analysts liked his blunt style, seeing it as a reflection of a practical approach to business that was also responsible for delivering production growth. That said, Inglis needed to cultivate a more positive external image as he saw himself as the next CEO of BP. Assuming Hayward stuck to his promise of staying for just five years, Inglis would be only 52 years old when Hayward stood down; as head of E&P, he had an edge over any other potential candidate.

The presentations covered all of E&P but focused heavily on the Gulf of Mexico. They delivered two clear messages. The first was on 'efficiency'. BP's platforms were shown to be operating ever closer to their design capacity. The company presented graphs showing that, while higher steel prices, dearer rig rental rates and higher wages across the industry were driving up production costs for rivals Chevron, Total, Exxon and Shell, BP's costs had been falling consistently since 2007. BP was also finding new barrels at a lower cost than anyone else. Mike Daly, the head of exploration, said that, between 2002 and 2007, BP had found over 5 billion barrels of oil and gas at a cost of just over $8 billion.

Chevron and Exxon had spent around the same amount each but only managed 7 billion barrels between them. Shell fared even worse. Safety got a look in, and no one objected when the statistics only referenced the personal safety measure of 'recordable injury frequency'.

The second main message from the presentations was 'technology'. BP claimed to be the technological leader in its field and highlighted developments such as its seismic technology and its ability to crunch geological data better than others.

After the presentations, analysts were taken around the main building and shown some of BP's high-tech gadgetry, including facilities from which BP could remotely monitor wells in real time, thanks to a 1,000-kilometre fibre-optic cable that ran from Pascagoula, Mississippi to Freeport, Texas and connected all BP's Gulf platforms in between. The analysts thought it a wise move to keep an eye on what was going on inside those holes in the sea floor. On the second day, the analysts were taken on a visit to the *Thunder Horse* platform and to the Texas City refinery.

The visiting analysts were mostly men, mostly in their thirties and forties, and they were predictably impressed with the big boys' toys they saw. When they went home, mostly to London, they duly wrote a series of glowing research notes about the field trip. The message on costs had been received loud and clear. Citigroup said it saw 'a clear and changed corporate culture and cost focus' and affirmed its 'buy' rating on the stock. It acknowledged that Shell, too, was talking about cutting costs but said it was worried by the Dutch company's 'more conservative messaging, if not delivery, around cost savings'. Morgan Stanley also noted that 'The strategy is to be the low-cost producer in the best postal codes in the US.' The report did not question how BP achieved lower costs than its rivals. Given what it had seen in Houston, Morgan Stanley opined that BP's shares were 'inexpensive'.

The technology message was also noted. Another brokerage declared that 'there is no spin in what BP is saying . . . the Gulf of Mexico now represents operational excellence and its best practices are now going to be applied across the rest of the group.'[7]

The bullish notes circulated through the market and the financial press, adding to a perception that BP was ahead of rivals operationally and technologically. But the analysts' amazement was in part driven by ignorance. What they had seen was not as impressive as they thought – and what they had not seen should have worried them even more.

When BP showed off its facilities for monitoring wells in real time, it did not point out that, when it came to drilling exploration wells, the most dangerous kind of well, BP used an external contractor. Nor did it point out that this external contractor only provided the service from nine in the morning until five in the afternoon, even though exploration drilling was a 24-hour activity. Had the analysts been shown around Shell or Exxon's offices they would have seen in-house facilities for real-time monitoring of all wells, including exploration wells.[8] Another thing the analysts might have been shown at Shell or Chevron was a cement laboratory, where the companies could test cement recipes recommended by their specialist cementing contractors. BP didn't have a lab for such purposes, and instead relied entirely on the cementing contractor to deliver good, finished mixtures.

The fact was that the visiting analysts had seen a second-tier operator, who just happened to be the biggest and brashest. Indeed, the fact that BP admitted spending half as much on research and development as rivals Exxon or Shell did should have led analysts to question how BP had achieved its claimed technological supremacy.

BP's peers were under no illusions about its technical skills. Shell had stopped partnering with the company on new projects out of disdain for its abilities. And BP was well aware of its rivals' unease regarding its practices. In early summer 2010, at the height of the oil spill, a group of around 20 BP executives and PR advisers was convened to discuss the Herculean task of rebuilding the company's reputation. According to someone present, towards the end of the four-hour brainstorming session a plain-speaking executive stood up and declared the problem wasn't one of reputation but of reality: 'If you asked anyone from Exxon or our other rivals what they thought of us,' he said, 'they would all say the same thing: we push things too far.'

MC-252

On 19 March 2008, the Minerals Management Service held lease sale 206. The agency, which was both the licensing authority for, and the regulator of, the offshore oil industry, offered 615 tracts on more than 28.5 million acres in the central Gulf of Mexico for bidding. Mardi Gras was more than a month past but a sense of carnival hung in the air as executives from dozens of oil companies flocked to the New Orleans Superdome for the auction. Oil prices were at $100 a barrel, making President Bush even more eager to see an expansion of domestic oil production. Dirk Kempthorne, Secretary of the Department of the Interior, which oversaw the MMS, had flown in from Washington to officiate at the proceedings and a group of Iraqi officials were also in town, to pick up tips on attracting oil investment.

The oil companies did not disappoint. They had brought their chequebooks and together would bid a total of $3.7 billion for the licences on offer: the highest amount raised in 54 years of federal offshore auctions, and comfortably beating the previous high of $3.4 billion achieved at an auction in 1983. 'Today MMS won the championship,' a beaming Kempthorne declared.

BP loomed large at the auction, bidding for more licences than any other company. If it wanted to retain its leadership position, it had to continually fill its 'hopper' with new opportunities – not least because of the high depletion rates suffered by deepwater wells in the Gulf. Production at every oil field starts to fall soon after oil begins to flow, because no more oil is being produced down below, and while traditional fields decline at 5 to 8 per cent per annum, deepwater Gulf of Mexico fields could decline at several times the global average. BP needed to push on and drill more wells, deeper and faster, than it had been used to doing in the past.

So it was good fortune that it enjoyed success at the Superdome in 2008. One of the licences it picked up was Mississippi Canyon block 252, or MC-252 for short. Internally, the company had nicknamed the prospect 'Macondo'. A group of Colombian workers had won a charity auction and decided to honour their great writer and countryman Gabriel García Márquez by naming the prospect after a town that appeared in his novel *One Hundred Years of Solitude*.[9] At $34 million, it wasn't one of the most valuable leases sold that day, reflecting the fact that the companies bidding thought it was not likely to hold the biggest reserves. Consequently, as BP planned its first well over the coming two years, the prospect attracted little attention.

BP sold a quarter share in the licence to Anadarko Petroleum, a Houston-based oil company that was active in the Gulf and making a name for itself as a savvy explorer thanks to big finds in West Africa. BP also sold another 10 per cent share to MOEX, a unit of Japanese trading house Mitsui. Such 'farm-outs' are a regular part of the exploration process. With individual wells costing $100 million or more to drill, even large companies like BP like to spread the financial risk involved. Anadarko and Mitsui would contribute towards the drilling costs on the basis on their stakeholding, and enjoy a commensurate share of the profits. Similarly, if anything went wrong, Anadarko was liable for 25 per cent of the cost and MOEX for 10 per cent.

When it came to devising a plan to drill MC-252, BP made an unusual decision. Oil wells are constructed by drilling through the earth's crust. As the drill bit bores down, it travels through layers of different types of rock. Some of these will be porous, some not. Some will contain water and some will contain oil or gas. Some of the areas of oil and gas will be large enough to exploit and most others will not and instead represent a hazard and nuisance. Naturally, one doesn't want these layers to commingle. On land, this could lead to contamination of the water table; under the sea, it runs the risk of oil seeping into porous layers rather than up to the rig. Hence, the well bore must be sealed off from the surrounding rock structures. To achieve this, the well is lined with steel piping that is cemented into place to ensure no gas or oil can escape up the outside. This lining of the well is done in stages, with the use of gradually narrower casing pipes as one goes deeper. For added safety, a second set of steel piping is run inside. This inner casing, known as the production casing, can either be a single pipe that runs all the way from the well head to the reservoir, or a two-piece system. The first system is known as a 'long string' and the latter, more complex system, known as a 'liner and tie-back' system.

BP chose the 'long string' system, for a number of reasons. First, it was the cheapest to install. It was also likely to be the cheapest to maintain over the lifetime of the well. The company would later argue[10] that it also had the advantage of being better at preventing a build-up of pressure between the outer well wall and the production casing, but this was not a view shared by most other oil companies. Indeed, the long string design was rarely used by BP's rivals, and was usually reserved for drilling into structures they had previously drilled and knew well.[11] Even though it was more expensive to install, companies like Chevron, Shell and Exxon tended to use the liner and tie-back system. The reason was risk. The liner and tie-back had an additional barrier between the reservoir and the surface, reducing the risk of leakage from the reservoir – the big worry for a driller. It was also easier to cement into place, again reducing the risk of leakage. Chevron, Shell and Exxon figured that the extra money they spent on the liner and tie-back system was like an insurance premium. BP took a different view. BP was comfortable with risk.

BP's decision to use a long string design for its exploration well was especially unusual given the reservoir they planned to tap was expected to have what is known in the industry as HPHT: high pressure and high temperature. HPHT fields are more challenging to develop than traditional fields. They could have oil at temperatures well above the boiling point of water, and even by the time the oil hits the surface it could be hot enough to burn workers' hands through the metal piping on the rig. An oil company wants a certain amount of pressure in a reservoir to drive oil up to the surface, but very high pressure requires more robust equipment so that pipes don't snap or valves fail. The complexities of developing HPHT fields is such that companies only began to tap them in the mid-1990s.

While BP's design was unusual, however, it did not break any MMS rules. The agency approved BP's plan in April 2009. The following October, the *Marianas*, a drilling rig owned by Transocean, a company that operated the world's largest drilling fleet and which provided drilling services to all the big oil companies, 'spudded' the Macondo well, as the breaking of ground is known in the industry. But the *Marianas* didn't get far. The arrival of Hurricane Ida a month later forced the evacuation of the rig. When the workers returned, they found the *Marianas* had been damaged and would need to return to port for repairs. In January, another Transocean rig, *Deepwater Horizon*, was towed to MC-252 block. *Deepwater Horizon* was one of the most advanced rigs in the world. It had been under contract to BP since it left the Korean shipyard that built it nine years earlier. The rig arrived at block MC-252 fresh from drilling one of the deepest wells ever drilled – at a depth of 35,000 feet – at the Tiber prospect. That well had found a 'giant' oil discovery. The expectations for MC-252 were more modest in terms of both depth and reserves.

Deepwater Horizon's top side sat on four legs connected to two pontoons. When the rig was in place, the pontoons sat beneath the water, giving the impression the rig was standing on four feet in the water. This semi-submersible mode of operation gave increased stability, and semi-submersibles had become the favoured rig type for drilling in the potentially choppy ultra-deepwater. But it wasn't long before *Deepwater Horizon* began to have problems.

Deeper and deeper

The nexus of the drilling process is the drill bit. Modern drillers favour 'roller cone bits', which have wheels with teeth at their tip for cutting through the rock. The first roller cone drill bit was invented by the father of the billionaire eccentric Howard Hughes, and was the basis of his fortune. A heavy fluid called 'drilling mud' is pumped down the hollow drilling pipe to lubricate the drill bit. The mud comes back up, bringing rock cuttings with it, via the gap between the drilling pipe and the well wall. The first problem the men on the drilling rig noticed while working on the Macondo well was that the amount of drilling mud they pumped down the well was not equal to the amount they got back out at the top. This phenomenon, known as 'lost returns', meant that they were losing fluid into the rock at the bottom. This was not good.

The whole point of exploration is that the drill bit will, at some point, hit a reservoir that contains a lot of oil and gas. This reservoir is not a hollow space, but rather a porous rock which is crammed full of oil and/or gas, which would seep to the surface were it not locked in by a non-porous cap rock above it. Since the oil and gas is under pressure, there is a risk that, as soon as one cracks through the cap rock, hydrocarbons will surge up the well. Naturally, this would risk an oil spill or gas explosion, but drilling mud prevents such an upsurge from happening. The mud column, possibly miles high, exerts an enormous downward pressure, keeping any hydrocarbons, which could be found in several different layers, at bay. The role of the drilling mud is so important that its pressure is carefully managed and monitored.

When the drilling mud at Macondo began to disappear into fractures in the rock, the driller became concerned that the overall pressure could drop, allowing hydrocarbons to seep into the well. Fractures can form for a number of reasons: the weight of the mud can damage delicate rock, for instance. Or rock can be fractured by drilling too quickly.

The crew on *Deepwater Horizon* dealt with the leakage of drilling mud by pumping in material to try to clog up the cracks in the rock. It worked and they continued. However, on 8 March, Macondo threw up another worrying surprise.

After just over a month of drilling, *Deepwater Horizon* suffered what is known as a 'kick', when gas seeped into the well and surged towards the surface with a force that shook the drill pipe.[12] The driller activated a 450-tonne piece of equipment known as a blowout preventer (BOP), a mile beneath the rig on the seabed, which was designed to seal off the well in such a situation. The BOP did its job, stopping the gas from reaching the surface and potentially igniting.

It wasn't clear what had caused the gas seep. Measurements showed that the pressure in the well bore had dropped beforehand – possibly due to mud escaping into fractures in the rock – and this could have created the opportunity for gas to creep in. These kicks often occur for no clear reason, and since rigs have multiple systems to ensure they do not lead to disaster, no one gets too worried about them. What was worrying about the 8 March 'well control incident', as these are known in oil industry lingo, was that it took workers 33 minutes to react to the ingress of gas. This was a remarkably slow reaction time. Had the volume of gas in the well been higher, the rig could have suffered a blowout.

The kick was so serious that it caused the drill pipe to get stuck in the well. BP had to sever the pipe and drill around the area where it was stuck. This added further delays to the already behind-schedule project. The kick was one of several events that led workers to describe Macondo as 'the well from hell' and 'a nightmare well'. In truth, the kick was a gift. It was a warning that BP's, and Transocean's, procedures needed to improve. Unfortunately, the gift was left unopened.

BP produced a report on the incident, which criticised the drilling crew's response time. Nonetheless, BP and Transocean staff on the rig were only given 'verbal feedback' of an undefined nature. They did not see a copy of the report.[13]

A month later, BP began to suffer further lost returns: drilling mud was escaping into the rock again. It took the usual remedial action: pumping in material to try to clog up the cracks. After drilling restarted, however, mud continued to be lost. BP was drilling slowly now, so excessive speed was not the problem. Quite simply, the rock was fragile and the pressure from the weight of the column of mud was cracking it. If the drillers persisted, the likelihood was that they would continue to suffer lost returns, and it would become impossible to control the pressure in the well. On 9 April 2010, BP accepted the Macondo well could be drilled no deeper.

The good news was that, although BP was over 1,000 feet from the depth it had initially intended to drill to, it had already hit hydrocarbons. Over the coming days, BP conducted tests to try and ascertain how much oil was in the reservoir. They knew they needed at least 50 million barrels of recoverable reserves for Macondo to be made into an economically viable producing well. Less than that and they would seal the well and move on.

The engineers decided there was comfortably enough oil to make Macondo economical. This left two jobs to do: run the production casing down to the reservoir, and pump in cement that would fix the casing in place and also seal off the reservoir ahead of the temporary abandonment of the well. That way, when BP returned to begin production, they could simply drill through the cement barrier at the bottom and let the oil flow.

But it wasn't simply a case of proceeding as originally planned. The long string had been BP's preferred option for completing the well, but the discovery of the fragile rock at the bottom forced a rethink.[14]

Cementing a well involves pumping cement down the steel casing so that it comes back up the gap between the casing and the rock wall. This requires a sufficiently wide gap between the casing and the well wall, and also sufficient pressure bearing down on the cement. To ensure the gap is wide enough for cement to flow, the production casing has to be perfectly centred within the well. This is more of a challenge with the long string because it is harder to centralise – in Macondo's case – an 18,000-foot pipe hanging down from the well head on the seabed than a pipe suspended from a collar halfway down the well, as the liner and tie-back system would have involved. The soft rock in the well increased BP's dilemma, because it would not be able to pump cement down with the level of pressure it would normally use. If the production casing was not perfectly centred, the chances of a botched cement job were high.

On 14 and 15 April, BP and engineers from its cementing contractor Halliburton ran computer programs that showed it would not be possible to safely install a long string production casing. BP's design team suggested switching to a liner and tie-back, which they calculated could be cemented safely, despite having to pump the cement at a low pressure. But senior BP drilling managers resisted the shift to the liner and tie-back scheme, which would take longer to install.

They called in BP's cementing expert. He duly ran the tests again, but this time he changed some of the assumptions that Halliburton had made.[15] Suddenly, the model churned out a different result: the long string option could, after all, be safely used. But the conclusion rested on a number of caveats: BP would have to pump less cement than its rules usually dictated, and it would have to pump the cement at a slower pace than it normally would. This would reduce the risk of cracking the fragile rock but it would also make it harder to ensure cement fully covered the gap between the casing and the well wall. Furthermore, the model could only be made to work on the basis that BP used cement that had been pumped with nitrogen gas to reduce its density. The weight of normal cement might have fractured the rock. The problem with such cement is that it is prone to channelling – when the bubbles in the cement come together to form routes through which gas can escape.

Drilling is a game of constantly overcoming the challenges that geology throws at you. This often means taking on new risks. The key to drilling safely, however, is to balance every new risk with an additional safeguard. For example, if Chevron or Shell decide to use a long string well, they might also install extra barriers in the well so that, should the cement job at the bottom not work, the gas will not go straight to the surface.

This risk-based approach to drilling is enshrined in law in the UK, with companies operating there obliged to set out a full risk assessment, known as a 'safety case', which outlines how they planned to manage dangers and work safely. Companies such as Shell and Norway's Statoil have in the past also completed safety cases when drilling wells in the Gulf of Mexico, even though they were not required to do so by US law and even though it increased cost, because they thought it prudent. Although used to completing safety cases in the North Sea, BP did not do so in the Gulf of Mexico, although it denied this increased risk.[16] Even in situations where BP's policies for operating in the Gulf of Mexico said it should conduct risk assessments, the company did not.[17] Consequently, no additional safeguards were taken to balance the additional risks. And the risks kept on mounting.

Halliburton recommended BP use 21 centralisers to ensure the drill pipe was fixed into place. But John Guide, the Houston-based head of the Macondo well team, told the drillers to proceed with only six because that was how many there were on the rig. Centralisers were cheap but the time it would take to get them was not: the rig was costing over $1 million a day to hire and run. Brett Cocales, who worked for Guide, disagreed with this decision but said to a colleague in an email on 16 April: 'Who cares, it's done, end of story, [we] will probably be fine and we'll get a good cement job.'

On 18 April, shortly before cementing of the well was due to proceed, a Halliburton engineer sent an email to several BP and Halliburton personnel warning that the latest model he had run on the OptiCem™ program predicted the existing cementing plan had a 'severe' gas flow potential. BP engineers had a look at the calculation, decided it contained some errors, and disregarded it. Without re-running the model using the correct parameters, they prepared to do the cement job. But not before making one last break from industry tradition. Guidelines laid down by the American Petroleum Institute, the industry lobby body, recommended that companies fully circulate the mud in a well before doing a cement job. Doing this allowed drillers to test the mud at the bottom of the well for signs that gas was seeping into the well. BP did not do this, choosing instead to circulate the mud for only a fraction of the recommended time.

After the cement job was finished, Halliburton's cementing specialists flew by helicopter from the rig to the BP heliport at Houma, Louisiana. BP had considered having them stay longer, to run a test called a cement bond log, or CBL. The CBL was a sure-fire way to test whether the cement job had worked. But BP decided against running a CBL. It was not a regulatory requirement and would only cost more time and money.

BP had avoided expense but missed yet another opportunity to avoid disaster.

Disaster

In spite of the much-vaunted roll-out of the operational management system (OMS), investigators would later slam BP's practices. But there was one area in which the company's systems matched those of any rival: its benchmarking of drilling costs. BP carefully tracked the time its drillers around the world took to complete different tasks. The quickest times for drilling 10,000 feet of well, installing casing or cementing a well were recorded and then used to construct a time estimate for drilling future wells. At the end of every day on *Deepwater Horizon*, the time it had taken to complete a given task was compared against the 'Best of the Best' data and the results shared with the rig crew. BP had a 'risk register' to which members of the Macondo team were invited to contribute ideas on potential risks that could be faced in drilling the well.[18] But the register was for matters that could cause delays or boost costs, not increase safety risks.

BP was also meticulous about monitoring cash spend. Every dollar spent, from $15 on a cargo box to $533,000 for the daily rental of the rig, was recorded. BP engineers had their bonuses and promotional prospects linked to drilling efficiency, so their incentive was to press ahead quickly and to spend as little as possible. Conversely, Transocean was being paid by the day, so had reason to take all the time in the world. Similarly, Halliburton had every incentive to conduct as many cement bond logs or other tests for BP as could be imagined. Yet BP's constant updates on how efficiently the drilling was proceeding versus BP's ambitious benchmarks would have had a predictable impact on the contractor staff: BP was one of Transocean and Halliburton's biggest clients and was not be antagonised.

By 20 April, Macondo was 45 days behind schedule and BP's wells team had calculated it would come in at $58 million above its original $96 million estimate. If staff felt pressure to finish the well, they were also keen at a personal level to put the difficult project behind them. The constant changing of drilling plans had been particularly frustrating for the Transocean staff and the BP well site leaders (WSLs), the BP representatives on the rig who oversaw the drilling.

John Guide, the leader of the Macondo programme, had on 17 April written an email to his boss, David Sims, outlining these frustrations: 'David, over the past four days there has been so many last minute changes to the operation that the WSL's have finally come to their wits end. The quote is – "flying by the seat of our pants".' He noted that 'the insanity' was even getting to staff on the Macondo team in Houston, including Brian Morel, a junior drilling engineer on the team that designed the well. 'This morning Brian called me and asked my advice about exploring other opportunities both inside and outside of the company.' He signed off presciently: 'The operation is not going to succeed if we continue in this manner.' If Sims was concerned by this, the opening line of his response didn't show it: 'John, I've got to go to dance practice in a few minutes. Let's talk this afternoon.' He urged Guide to remain positive before signing off: 'I'll be back soon and we can talk, we're dancing to the Village People!' Sims later said he had interpreted the comments as a temporary letting-off of steam rather than a sign of any fundamental concerns.

And so it was, in this atmosphere of high cost and time pressure, that the workers pressed ahead. The drilling plan envisaged temporarily sealing off the well with a concrete plug near the top once BP was happy the concrete barrier at the bottom was solid. On 20 April, BP and Transocean staff prepared to undertake their one test to ensure that the cement job at the bottom of the well had successfully sealed off the well bore from the reservoir.

Since the well was still full of mud, it was impossible to know whether it was the mud or the cement barrier at the bottom of the well that was keeping the hydrocarbons in the reservoir at bay. Since BP planned to replace the mud with lighter seawater before setting the upper cement plug, it was crucial that the cement barrier below was up to the job. The test to ensure that the cement was good involved reducing the pressure from the mud, slightly, and seeing what happened. The drillers had the mud perfectly balanced, so even a small reduction in the pressure would create an 'underbalanced' situation, which would allow gas to flow into the well through any hole in the cement. If the pressure rose again, it would be clear that oil or gas was entering from below – in which case, drilling mud would be pumped back into the well bore, increasing downward pressure and pushing the hydrocarbons back into the reservoir until the cement job could be remedied. Problems with cement jobs were common and cementing specialists had an array of tools to conduct repairs.

The negative pressure test was conducted over a period of five hours from around 3 p.m. on 20 April. The test repeatedly showed that pressure was building up in the well, but the dayside BP 'company man' on the rig, Bob Kaluza, and his Transocean colleagues misread the signals. The testing continued after 6 p.m., when Kaluza was replaced by Donald Vidrine, BP's night-time well site leader. Vidrine also failed to correctly read the signs the negative pressure tests were giving him. That neither man knew how to read a negative pressure test correctly was a staggering indictment of the quality of training of BP staff. Yet crucially, the Transocean staff also failed to spot the danger and even offered innocuous explanations as to why the pressure in the well was high and rising when it should have been flat. Despite its critical importance, neither company had given its staff formal training in how to read a negative pressure test.

Having convinced themselves that the cement had successfully isolated the reservoir, the men continued to take heavy mud out of the well and replace it with lighter seawater. But the cement at the bottom of the well had not formed a seal. With no other barriers between the reservoir and the rig – most other big oil companies would have included additional barriers in their well designs – the weight of the mud was the only thing keeping Macondo at bay. The men on the rig were unwittingly creating a clear channel between themselves and the reservoir.

Out of hours

Gas continued to fill the well; the pressure inside continued to rise. Sitting in the drill shack, head driller Dewey Revette was experienced enough to know what rising pressure in the pipe meant. But he did nothing about it. People would later spend tens of millions of dollars trying to find out how he missed the sign of imminent catastrophe, but it was probably because he was human. Down in the mudlogger's office, the rising pressure on the drill pipe was being displayed on one of around a dozen screens. But the mudlogger, whose main job was to monitor the flow of mud in and out of the well, was taking a break, as humans occasionally need to do, during a 12-hour shift.

Exxon and Shell knew offshore workers could easily be distracted aboard a busy rig. Both companies had computer-filled centres onshore that automatically monitored the pressure in the wells, 24 hours a day, ready to flag up any emerging dangers. BP also had a monitoring centre – outsourced, of course – but it only operated during office hours. Had Revette taken his eye off the pipe pressure gauge four hours earlier, just before 5 p.m., he might have been alerted by a call from shore telling him he was minutes away from disaster. As it was, at 9 p.m., the computer banks onshore were switched off and the people who might have made the call had gone home for the day.[19]

It wasn't until 9.40 p.m., when drilling mud began to spew from the well onto the rig floor, that Revette realised his life and the lives of everyone on the rig were in peril. He immediately flipped a switch to activate the rig's blowout preventer (BOP), the 450-tonne block of valves on the sea floor designed for such a crisis. But it was too late. The gas was already above the BOP and was accelerating towards the rig as the pressure from above dropped. The force of it pushed back the mud that was designed to contain it, and a fountain of drilling mud shot into the air, splattering the supply ship *Damon B. Bankston* alongside the rig with what looked to its captain like black rain.

Down in the engine control room, chief mechanic Douglas Brown was sitting at his computer when he heard a hissing noise. Then a blue light began flashing on a control panel on the wall: an alarm notifying him that there was gas on the rig. This wasn't necessarily a sign of impending doom: drilling decks were 'intrinsically safe', designed to have no potential sources of sparks. Gas was only a problem if there was enough of it to get beneath the main deck to where the engines – which could provide a spark – were located.

It was not long before Brown heard the engines speed up and rev harder. The gas had entered the engine rooms and was being sucked in by the machines.

Seconds later the first blast shook the rig.

On deck, Revette and his colleagues in the drill shack were killed instantly. Elsewhere on the rig, computer monitors exploded, light bulbs popped and control panels sparked. Steel doors were ripped off their hinges and crumpled like tin cans. Ceilings collapsed. Some men were injured by flying debris; others were thrown against walls and knocked unconscious. Crane operator Dale Burkeen was engulfed in smoke and flames as he climbed down from his cab.

A second blast erupted. Burkeen was thrown to the deck.

Chief engineer Steve Bertone had been in bed when the first blast ripped through the rig. He threw on his clothes and rushed up onto the bridge. He looked through the window to see the derrick engulfed in flames. Inside, the scene was no better. There was no power; the engines were not working and neither were the telephones. There was only one way to save the rig now: activating the emergency disconnect system (EDS). This would slice through the drill pipe, seal the well, disconnect the rig and leave *Deepwater Horizon* free to float away.

But the EDS was seen as the nuclear option in a well control situation. It would take days or even weeks to reconnect a rig once it had been severed. The cost could run to tens of millions of dollars, and Transocean had ruled that the EDS could not be activated without the highest-level approval. Not even the captain had the authority to activate the EDS. It was not until Jimmy Harrell, the offshore installation manager, the boss of *Deepwater Horizon*, had arrived on the bridge and given his approval that the switch was hit.

Bertone rushed down to the engine room to try to regain power, but to no avail. When he returned to the bridge, the flames on the deck told him the EDS had also not worked. The inferno was clearly still being fed from below.

The order to abandon ship was given. The men had done weekly drills on abandoning the rig; in theory, they should have been prepared. But the drills were always conducted on Sunday mornings, in good weather, and never involved rig staff actually entering the lifeboats and being lowered into the sea. Now, there was a scene of total chaos as men fought their way to the life vessels. When the first ones arrived, they were forced to wait for their colleagues to arrive under the glare of the burning derrick. Some decided against waiting and leapt into the Gulf of Mexico. Bertone and his colleagues on the bridge were unable to reach the muster points and instead boarded a life raft, which they lowered into the water on pulleys. The sea around them was on fire.

When the survivors had been hauled aboard the *Bankston*, the managers took a headcount: 115. Eleven of *Deepwater Horizon*'s crew were missing, presumed dead: Jason Anderson, Dale Burkeen, Donald Clark, Stephen Curtis, Roy Kemp, Gordon Jones, Karl Dale Kleppinger, Blair Manuel, Dewey Revette, Shane Roshto and Adam Weise.

6

PR Playbook

News of the explosion on *Deepwater Horizon* travelled quickly up the chain of command at BP. Shortly after 10 p.m., the company's control room in Houston received an email from the *Damon B. Bankston* alerting it to an explosion on the rig.[1] BP's Gulf of Mexico boss was quickly informed. He woke Doug Suttles, the chief operating officer of the exploration and production division, to tell him the news. Suttles informed his boss, Andy Inglis, who was in London at the time, and he had the task of telling Tony Hayward.[2] The CEO was jogging in Hyde Park when Inglis called, so he left a voicemail. At 7.24 a.m., three and a half hours after the blast, Hayward heard the message.

'What the hell did we do to deserve this?' he asked colleagues soon after hearing the news.[3]

Hayward convened a team of his top executives to discuss the tragedy. Among them was his new head of press, Andrew Gowers, formerly editor of the *Financial Times* and then head of communications and marketing at Lehman Brothers, which, like BP, he had joined at just the wrong time.

Gowers, Hayward and the other BP executives knew 11 men were missing from the rig. Over the coming months, every public statement made by politicians, BP executives or media commentators started with an expression of sympathy for these 11 men who would never be found, the uniformity of the statements reflecting the PR rule that one must always show more sympathy for human tragedy than for economic or environmental loss, even if one might be more concerned about the latter.

The executives looking at television pictures of *Deepwater Horizon* in flames from the safety of BP's London HQ doubtless felt genuine sorrow at the loss of colleagues, but what probably worried them as much was the prospect of what might happen when the rig stopped burning. If oil began leaking into the Gulf, BP had a major problem on its hands. Inglis, Hayward and BP's other top engineers knew of only one sure-fire way to stop a leaking well 5,000 feet below the sea. Unfortunately, it would take over three months to implement. If a leak was discovered, they would need to find a new solution, and fast.

Gowers' task was better defined. After all, the crisis-PR playbook was largely based on an oil spill: the 1989 *Exxon Valdez* spill off the coast of Alaska. The botched clean-up effort did little to cool public anger but what had really turned the name Exxon into a byword for uncaring capitalism was the way in which the company chose to communicate news of the accident. For one thing, it had sent lower-ranking officials to the site rather than its chief executive, which gave the impression that the company did not take the matter very seriously. When the CEO, Lawrence Rawl, did finally emerge to speak about the disaster, around a week after it had happened, his comments served only to inflame the common perception of an uncaring mega-corporation. In television interviews he proved unable to provide details of the clean-up plan, saying it wasn't his job to know such things. He blamed government officials for the botched response and the media for stirring up such public outrage. Forty thousand customers cut up their Exxon credit cards in response – which Rawl said he wasn't bothered about, as the cards weren't being used much anyway – and the company became a target for environmentalists worldwide. The spill and its handling had ramifications for the whole oil industry. President George H. W. Bush was forced to tighten industry regulations and shelve plans to open the Alaska National Wildlife Reserve to drilling – plans his son would also have to shelve a decade later, following BP's Alaskan oil spills.

The *Valdez* experience led to acceptance of a number of broad PR principles. First, the chief executive should fly quickly to the site of a disaster to emphasise that the corporation was concerned and planned to put all its efforts behind tackling the problem. Secondly, the company should strive to be open. Alongside Rawl's silence, Exxon had been parsimonious and inaccurate with the information it gave out to the media. Finally, it was critical that a company responded competently to any disaster, and was seen to do so.

In addition to such accepted wisdom, BP had its own contingencies for dealing with crises. It even had a group crisis handbook outlining the steps that should be taken. First among these was the formation of a steering committee, made up of the CEO, the PR director and other top officials, which would meet at least once a day to plan the response to the crisis.

All of this should have provided a seamless framework for shaping, developing and communicating a well-thought-out message. But Gowers had a problem, since Hayward appeared already to have made up his mind about what went wrong on the rig, and furthermore, that this conclusion would be BP's public position. By late morning on 21 April, as *Deepwater Horizon* still burned, BP had issued a press release. It was only 171 words long and seemed innocuous enough. But the release's title: 'BP Offers Full Support to Transocean after Drilling Rig Fire', was an indication of the calamitous decision that had already been taken at the very top of BP. The crisis was less than 12 hours old and Hayward had made his first big mistake: he had decided to blame the whole accident on Transocean. A website was set up to provide information on the accident. Again, the title said it all: 'Gulf of Mexico – Transocean Drilling Incident'. (Later, after the government got involved, Transocean's name would be removed from the website's strapline and replaced with BP's.)

In the coming weeks, Hayward would tell every reporter who would listen that Transocean had been in charge of the whole operation. BP was the architect, but Transocean was the builder executing the work. The fact that a problem had occurred was clearly the fault of the builder. After all, if one hired a builder to construct a new conservatory and in the process he knocked a hole in a wall, one would hardly blame the architect.

The denial of responsibility was not simply analogical. It was utterly specific. 'It wasn't our accident,' Hayward told me a week after the blast. 'This was not our drilling rig, it was not our equipment, it was not our people, our systems or our processes. This was Transocean's rig, their systems, their people, their equipment.' He singled out the failure of the blowout preventer as the primary cause of the accident. This 'fail-safe' piece of equipment was designed to prevent such accidents, and it was Transocean's responsibility. BP executives would later say Hayward genuinely believed that Transocean, and not BP, was to blame. The fact was that his comments were utterly wrong. BP knew that the person who ultimately called the shots on the rig was a BP representative, known in industry terms as 'the company man'. BP had also designed the well and chosen the processes, including the number of centralisers, the circulation time of the drilling mud and the type of cement to be used.

Almost instantaneously, BP had broken a key rule of handling public relations in a crisis: be demonstrably honest and transparent. The principle behind this policy was simple: in a crisis, a company would face intense media scrutiny; every claim would be checked and inaccuracies uncovered. Any attempts to hide the truth would only play into the hands of the company's critics, and nothing else the company said would ever be believed again, all of which would result in a more damaging crisis and a longer, harder route to recovery. BP didn't yet know the full extent of the crisis, but already Hayward was defending the company with unsustainable claims.[4]

Two days after the rig blast, *Deepwater Horizon* sank, severing the reservoir's connection to the surface. With no outlet for being burnt off, there was only one place for any escaping hydrocarbons to go. BP readied dozens of vessels for potential oil-skimming duty, gathered hundreds of thousands of feet of absorbent booms and tens of thousands of gallons of oil dispersant, and booted up computer programs that could model potential trajectories of oil spills. Hayward flew to America.

In Houston, his team was surveying the seabed with miniature submarines known as remotely operated vehicles (ROVs). The riser, the long pipe that had connected the rig to the blowout preventer, had snapped in two, and while part of it remained connected to the rig, the other end, still attached to the blowout preventer, had twisted and fallen to the seabed. To everyone's immense joy, the ROVs showed no signs of leakage around the blowout preventer. But the relief was short-lived. When the ROVs travelled the length of the riser and reached the end, they captured footage that alarmed the men sitting in BP's Houston control room: oil, gushing out at a ferocious rate. Three days in and BP had an oil spill. And it was the worst possible kind of spill: ongoing, in a location that made it almost impossible to tackle. This was the nightmare scenario for any oil company.

By now, the US government had established the 'Unified Command', made up of state agencies and BP staff. It was officially headed by the Coast Guard, but BP called the shots. BP's engineers watched the ROV footage and told the government that oil was escaping at a rate of 1,000 barrels per day. The company never disclosed how this estimate was reached, but it shouldn't have been a guess: BP had many computer models at its disposal that could calculate the flow rate based on the size of the plume and the size of the pipe. It also had an idea of the flow rates the reservoir was capable of if the well head were removed – up to 100,000 barrels per day, by their estimation.[5] Indeed, it was the existence of such established knowledge on judging well-flow rates that allowed BP to appear credible when it gave its estimate.

On 24 April, the day after Hayward arrived in Houston, the 1,000-barrels-per-day flow rate was released via the Unified Command's information centre, thanks to which BP could later claim that the estimate was an official government one rather than its own. Neither the estimate nor the attempt to blame the government for compiling it would stand up to scrutiny. BP had made its second major communications error.

So what's the plan?

Deepwater Horizon took some time to take off as a media story. News of the initial blast did not provoke outrage. Even the discovery of the leak failed to incite a media frenzy. Ali Velshi, an anchor on CNN, which later provided excoriating coverage of the spill, noted on 27 April that 1,750 barrels of oil leaked naturally from the ocean floor on a daily basis. 'That's normal leakage and that's why we've got those microbes,' he said. 'They can handle a certain amount of oil.'

Underpinning the initially restrained media response was an expectation – backed up by BP – that the spill would not last long. But the optimism on all sides was short-lived. A day after the 1,000-barrels-per-day estimate was issued, government scientists realised it was hopelessly low. A scientist with the National Oceanic and Atmospheric Administration (NOAA), the federal agency that oversees fisheries and the condition of the atmosphere, calculated a flow rate of 5,000–10,000 barrels per day. BP objected to the higher estimate and suggested a rate of 1,000–5,000 barrels per day. In the end, the Coast Guard accepted the BP figure. After all, who knew more about oil spills: an oil man or an oceanographer?

Nonetheless, the release of the higher flow rate and the delay in sealing the well escalated the crisis significantly. Homeland Security Secretary

Janet Napolitano and Interior Secretary Ken Salazar announced a joint investigation into the blast. A number of Congressional Committees also announced probes.[6] Louisiana Governor Bobby Jindal declared a state of emergency and mobilised National Guard troops. Meanwhile, the higher flow rate knocked another 7 per cent off BP's shares,[7] which had suffered only slightly from news of the explosion. Shares in Transocean, well partner Anadarko Petroleum, well cementer Halliburton and blowout preventer manufacturer Cameron International also suffered big drops.

A week after the blast, the media adopted a much tougher tone. TV anchors began talking of an 'enormous environmental disaster'[8] and predicted that 'millions of barrels of oil are headed toward Gulf coast beaches'.[9] Journalists started referring back to Texas City and the Alaskan oil spills, implying that this disaster could be a result of further BP corner-cutting.

With pressure on BP growing, Hayward decided to take a bold move and fight back with the company's strongest weapon: money. He flew back to London to put in place changes that would allow BP to run without him while he led the response effort in America.

Coincidentally, he and I had a long-planned lunch appointment in London, which I fully expected him to cancel, but Gowers called to confirm it would go ahead. Indeed, while he had initially agreed to the meeting on the basis that the conversation would be off the record, Gowers now told me that Hayward would be happy to speak on the record. Clearly, I assumed, he had a message to deliver.

I had known Hayward for several years and he wasn't usually troubled by my presence, but when I met him in the lobby of the Reuters Canary Wharf building on 30 April, he was clearly, and perhaps predictably, uncomfortable. He was wearing a jacket and tie for perhaps the last time in over a month. I tried to break the ice by asking if he had been busy trying to keep out of his engineers' way. I knew that one of the pieces of advice often doled out to people in charge at times of crisis was not to micromanage. Hayward gave me a look of bemusement. 'No, I'm the general,' he said, 'and when you're the general, you have to lead from the front.'

I didn't realise it fully at the time, but this constituted Hayward's third big PR mistake: he had decided to front the response effort himself. If not the most fatal, it was certainly the most public of his mistakes in handling the crisis. As the CEO of a rival would later tell him, 'You stopped being the CEO and slipped into chief operating officer mode.'

On the way to the dining room, I tried again to lighten the atmosphere by asking about the name of the leaking well. I mentioned that the day before the rig blast I had begun reading Gabriel García Márquez's novella *Leaf Storm*, which, like *One Hundred Years of Solitude*, is set in a fictional town called Macondo. I asked if that was the source of the name. Hayward shrugged and said he didn't know. I wasn't surprised. It would probably be a bad sign if the CEO of a big oil company knew the etymology of the name of every well in his portfolio. What was worrying, however, was that he couldn't leave it at that. 'Maybe there were some Venezuelans on the rig,' he added, a little unsurely. I thought it undiplomatic to point out that García Márquez was not Venezuelan but Colombian. It wasn't important that Hayward didn't know the writer's nationality. I had only selected *Leaf Storm* to read because it was the thinnest in a pile of Márquez novels I had been given as a gift. But what Hayward's comment showed was that he didn't know when to stop speaking. Or, as an aide later put it to me, 'He doesn't stop short and he doesn't like to say "I don't know."'

As we prepared to sit down for lunch, I asked Hayward about his strategy for handling the public relations side of the disaster. He declared that BP would be judged by its response to the crisis. While, on the face of it, this seemed like an eminently reasonable approach, it was nothing of the sort. It struck me as arrogant and naive to expect that the company should be judged on its own terms, not on whether it caused the disaster or whether it was honest in its communications after the fact. It was to suggest that the situation should be one of BP opening its chequebook to the gratitude of us all.

But I didn't say this to Hayward. Instead, I suggested that relying on a good response effort to preserve BP's reputation might be a naive strategy. I explained the case of the McDonnell Douglas DC-10, a model of plane that was doomed to failure almost as soon as it went into production after one jet lost its engine and was sent into a tailspin. The incident was caught on camera by a tourist and the photos were printed in newspapers and shown on television globally. A subsequent investigation blamed poor maintenance but by this stage the damage was done: passengers believed the DC-10 had a design flaw and refused to fly in it, so airlines refused to buy it. Hayward didn't think this bore any relation to his situation.

The mood grew increasingly convivial, so I ventured another comparison and asked whether he aimed to replicate Johnson & Johnson's media success in handling the Tylenol poisoning scare in 1982. The drug-maker's quick and open response in pulling the product off the shelves when there was just a suspicion that its products had been tampered with became the gold standard of how a corporation can retain public trust by sacrificing short-term profits. But Hayward didn't think that case had any parallels either.

Not being an expert in crisis PR, I wasn't disappointed that I couldn't hit upon the tactic that BP was hoping to deploy. What was surprising was that Hayward could offer no analogy for success beyond 'We will be judged on our response'. Then again, he thought he had an ace up his sleeve. As we sat down to our lunch of pan-fried salmon, Hayward unveiled his bombshell. In what were obviously pre-prepared remarks, he declared: 'We are taking full responsibility for the spill and we will clean it up, and where people can present legitimate claims for damages we will honour them.'

The law required BP to pay whatever it cost to clean up the spill but when it came to compensating people who suffered economic loss – such as fishermen and hotels – BP enjoyed the protection of a $75 million legal cap on liability. Hayward said he would not seek to hide behind this: if anyone suffered economic loss and could prove it, BP would pay up.

Immediately after lunch we put the comments out on the wire. Within minutes the news was appearing on TV screens across the world. Hayward's declaration won BP goodwill in the US, especially among Republican politicians and some right-of-centre media outlets such as Fox News. It also convinced the British public and politicians that BP was trying to do the right thing. It served to underline that BP was being, in the words of Admiral Mary Landry, the head of the Unified Command at the time, a 'very responsible spiller'.

Nonetheless, the absence of a formed strategy for dealing with the upcoming media storm was about to outweigh all BP's largesse. A few days later, I wrote a story entitled 'BP chief an ops man who needs to show PR skills'. I don't think Hayward ever read it.

Crisis PR

It was clear early on in the crisis that BP would need some support in handling the PR campaign in America. The company's choice, Brunswick Group, came as a surprise to many in the PR industry. Brunswick was led by founder Alan Parker, widely considered one of the best-connected men in the UK. He holidayed with David Cameron, Gordon Brown was godfather to his son, and everybody took his calls. Brunswick was the UK's largest financial PR group, and Parker's half-share was estimated to be worth around £90 million.[10]

Brunswick had established a reputation as the go-to agency if one wanted to manage press coverage of mergers and acquisitions and other big corporate deals. But the firm was not known for crisis PR. Its name didn't ring many bells in America. The agency insisted that its US representative, Michele Davis, a former Treasury Department official, had contacts with the White House, but Brunswick did not have the large staff and wide US media contacts enjoyed by the kind of PR agencies usually hired by big oil companies.

The decision to hire Brunswick was Hayward's. As with many shy people, he liked to deal with people with whom he was familiar, and he knew Parker from his days as a non-executive director on the board of steelmaker Corus Group. Brunswick had advised Corus during a takeover battle that saw it swallowed up by India's Tata Group. For Brunswick, the oil spill represented a massive payday. It would generate tens of millions of dollars in fees and had the potential to establish the agency Stateside.

As early efforts to contain the leaking oil via the cofferdam steel dome and Top Hat mechanisms failed, the BP oil spill became the only story in America. In late April, Goldman Sachs had been the US corporate villain, with the Securities and Exchange Commission alleging the investment bank had committed fraud and misled investors. But within days, BP had rescued Goldman from the front pages. It is doubtful that BP could have done anything to avoid taking the place of the investment banks in the pantheon of public hatred, but its next few PR moves certainly accelerated the process.

Even before his famous gaffes, Americans were unmoved by images of Hayward on TV, quoting his idol Churchill and looking crestfallen as he surveyed soiled beaches. Exposing himself to such scrutiny was exhausting, but also pointless. No matter how sorry he looked, or how much he emphasised BP's good intentions, this could not trump the sight of brown scum on Gulf beaches. Compounding matters was the US administration's attempt to deflect growing public anger by referring to BP as 'British Petroleum'. Reviving the name BP had dropped 12 years earlier played on Americans' sense of being attacked from outside, and Hayward's English accent only reinforced the idea. While the PR playbook made it clear that Hayward should make an early appearance in the Gulf region, it wasn't necessary for him to stay there.

BP's rivals were baffled by Hayward's lengthy turn in the limelight. 'Tony should have been in Washington,' said the CEO of a rival oil giant, 'and not standing on the beaches, trying to calm down the fishes.'

Three weeks into the crisis, Hayward came up with a plan for a stand-alone body that would oversee the restoration of the Gulf once the spill had been stopped. He even decided who should lead it: Bob Dudley, director for the Americas. Dudley had been born in New York and brought up in Mississippi, so his accent fit, but more importantly his controlled manner was deemed perfect for such a highly charged situation. A number of people advised Hayward to implement the plan immediately, but he insisted that he should be the main face of the response effort until the well was capped. Supported by Brunswick, he continued to appear daily on US TV.

How much oil?

The other plank in BP's PR strategy, in so far as one had been conceived, was aggressively defending the 5,000-barrels-per-day flow-rate figure. Within days of this estimate being released, independent scientists began to propose much higher figures.[11] With BP's honesty being called into question, the company's response was short-sighted in the extreme.

Doug Suttles, chief operating officer of the exploration and production unit, was put on any television channel that would take him to defend what BP called the 'official' flow rate. And Suttles had all the qualities needed for the task. He had a southern-states accent, a clean-cut look, a folksy charm and the ability to appear incapable of hoodwinking people. But best of all, he had prior experience in defending the unpalatable to the public.

Before becoming Andy Inglis's number two, Suttles had run BP's Alaskan operations. His time there coincided with then-Governor Sarah Palin's move to increase taxes on oil companies. Oil prices had broken through $100 a barrel and Palin figured Alaska was due a bigger share of the windfall gains the companies were making. Naturally, the oil companies didn't like this. As the biggest operator in Alaska, BP took a leading role in opposing the hike. Suttles became a regular on Fox News and National Public Radio, in local newspapers and at conferences, talking about the need for domestic energy supplies to tame higher oil prices, and how the Alaskan tax hike would damage this goal. He said his team had looked at Palin's tax plan and concluded: 'By our calculations, Prudhoe Bay has the highest production tax now in the world.'[12] This was stretching credulity. ConocoPhillips calculated Alaska's tax rate on profits above around $100 a barrel would be 75 per cent. In Russia, where BP was the largest foreign oil producer and presumably knew what the tax rates were, marginal taxes of 90 per cent applied on every dollar of profit above $25 a barrel.

Similarly, Suttles had complained about the way Alaska's tax regime kept changing, creating an uncertain environment not conducive to investment. If America's approach to taxing oil companies was variable, it was in no small part because Washington spent a lot of time changing the rules to suit the companies themselves, especially when market conditions became harsh. The Minerals Management Service (MMS) had boosted royalty relief twice in five months in 1999, after oil prices hit $10 a barrel.[13]

Suttles' apparent eagerness to depict the flow rate as being 1,000–5,000 barrels per day – he did so at least three times on 29 April alone[14] – was especially questionable since, a day after the Unified Command issued the higher rate estimate, BP had devised an internal estimate of 1,063–14,266 barrels, calculations later discovered in internal documents by federal investigators.[15]

Throughout May, as doubts around the 'official' flow rate grew, Suttles and other BP executives continued to defend the 5,000-barrel figure, putting responsibility for it squarely on the US government. Bob Dudley went so far as to say, in an interview on MSNBC, that scientists who claimed Macondo was gushing up to 70,000 barrels per day were 'scaremongering'.

It was impossible, he said, to know for sure what the exact flow rate was, but 5,000 barrels was the 'best estimate'. In late May, Suttles told New Orleans paper the *Times-Picayune* that attempting to measure the flow was 'extraordinarily imprecise and we took a view very early on that we didn't think you could do it and we didn't think it was relevant either'.

The executives' professed disinterest in the flow rate was at odds with BP's own Oil Spill Response Plan for the Gulf of Mexico, which listed 'Determine size and volume of the spill' as 'critical to initiating and sustaining an effective response'. It went on: 'When a spill has been verified and located, the priority issue will be to estimate and report the volume and measurements of the spill as soon as possible.' The plan even outlined methods for assessing the size of the spill: 'Spill measurements will primarily be estimated by using coordinates, pictures, drawings, and other information received from helicopter or fixed wing overflights.' This Spill Response Plan was publicly available, so the disparity between BP's written procedures and its public statements were clear for all to see. And when the government employed methods such as those outlined in BP's response plan, the figures they came up with for the size of the spill were massively higher than BP's: an estimated 62,000 barrels per day at the outset. People knew the potential fines BP faced hinged on the amount of oil spilt, and naturally interpreted the company's attempts to downplay the flow rate as an attempt to limit liability.

Politicians and media commentators accused BP of a 'cover-up'.[16] President Obama, who had also slammed BP's other main communications strategy – blaming Transocean – as a 'ridiculous spectacle', even hinted that the company might have little incentive to be honest: 'It is a legitimate concern to question whether BP's interests in being fully forthcoming about the extent of the damage aligns with the interest of the public.'[17]

BP had had ample time to devise a communications strategy around the flow rate: three days had passed between the blast and the discovery of the leak, and another day passed before the leak was disclosed at a press conference. At the very least, its first statements should have emphasised the possibility of flow rates well above 1,000 barrels per day, if indeed that was the company's best guess. Instead, it seems BP either deliberately lied about the possible flow rate or its PR strategy was incompetent. Indeed, if it lied it was also being incompetent, because the lie would obviously be found out, with inevitable consequences.

CEO gaffes

By the middle of May, less than a month into the crisis, Hayward was becoming increasingly incensed at US media coverage of the spill. As far as he was concerned, TV and newspaper coverage was turning a difficult situation into a worst-case scenario.

After spending weeks trying to ingratiate himself to the American public, Hayward gave a series of interviews to the UK media in which he outlined his true feelings. He told Sky News that 'everything we can see at the moment suggests that the overall environmental impact will be very, very modest', and the *Guardian* that 'the Gulf of Mexico is a very big ocean. The amount of volume of oil and dispersant we are putting into it is tiny in relation to the total water volume.' It appeared that Hayward thought his countrymen might understand him better and be more sympathetic to his point of view. It also appeared that he thought people in the US had no access to UK media. In the weeks that followed, American TV anchors would begin their interviews with BP executives with reference to the 'tiny' and 'very, very modest' remarks.

BP's press officers tried to defend the comments by saying they had been taken out of context and that, in fact, they were largely technically correct. The latter may have been true – there was still a lot more water than oil in the Gulf – but in high-level communications, the broad impression one gives is more important than the technical accuracy of one's comments. Hayward might as well have said of the dead men that they would have died sooner or later anyway.

By now, the oil spill totally dominated the news. TV news channels displayed counters in the corner of the screen showing the number of days the oil had been leaking. Fears of an outbreak of war on the Korean peninsula and renewed diplomatic tensions over Iran's nuclear programme were ignored so that anchors could focus on the oil spill.

Hayward's comments prompted anger across America. When the famously calm President Obama failed to show sufficient fury of his own, he became the subject of criticism even within his own party. Every evening on CNN, James Carville, aka 'the Ragin' Cajun', the political consultant credited with getting Bill Clinton elected president, would rant about BP and Hayward. Director Spike Lee also got in on the act, suggesting there was 'environmental racism' in the response effort. Over on Comedy Central, *The Colbert Report* featured a sketch in which, by way of instruction to Obama, host Stephen Colbert beat up a foam dummy made up to look like Tony Hayward. Protesters picketed BP stations and offices around America, carrying placards of the green and yellow BP logo covered in splatters of black paint. In Louisiana, protesters trampled on the Union Flag.

BP's pain was increased by the fact people could see the spill on their TV screens at any hour of the day or night. Under pressure from Congress, BP was forced to broadcast the footage being captured by its ROVs, of the oil spewing from the end of the riser. 'As soon as we put up the footage of the plume, we were finished,' one senior adviser later said. 'It was like bleeding to death live on television.'

BP's public relations operation found itself hopelessly out of its depth. Its US press officers were sent out into the field, alongside representatives from Brunswick, to keep local leaders and citizens along the coast up to date on what BP was doing. PR officers were also flown in from St James's Square to man the Houston press office, although to call it that is being generous: it was a meeting room about 12 feet by 12 feet on the fourth floor of the main Westlake building, with a few tables pushed together in the centre and just enough room to seat half a dozen people around the edges. Three or four people worked in 12-hour shifts trying to answer incoming calls. There wasn't enough manpower to answer them all, so the press hotline was hooked up to an answering machine. It was probably as well. Many of the people calling were doing so only to vent anger or to threaten Hayward, especially in the early hours of the morning after the bars along the Gulf Coast had closed. Usually, if one needed a quick answer to a query, it was best to call press officers on their UK mobile phones. If they recognised your number, they might pick up.

The BP press team were used to dealing with financial journalists who wrote unemotional, carefully balanced stories, deep in context; they found themselves ill-equipped to handle the bare-knuckle tactics of 24-hour TV news and tabloid newspapers. One night, when a group of BP employees were out having dinner, Hayward went along towards the end to say hello. He sat down and a beer appeared in front of him. Gowers recounted this tale to me as evidence of not everyone in America hating the CEO. I suggested the beer might have been sent over by a tabloid photographer. The thought had not occurred to Gowers.

Hayward appeared similarly unaware of the potential for being captured on camera in situations that might be misconstrued. Indeed, a few weeks later a photo did emerge of Hayward, sitting at a bar beside an attractive young woman. It was taken on the Saturday of the England v. USA football match. He said he had been sitting at the counter of a restaurant, having lunch and a cranberry juice, watching the game, when the barman snapped a photo. Hayward denied knowing the woman and there was no reason to believe that he did – but the photo soon appeared on the Internet alongside claims he had been caught cheating on his wife. Newspapers, doubtless aware that they could be successfully sued for defamation, chose not to run the photo, saving both Hayward and BP unnecessary embarrassment, but it was a near miss.

All the time, BP's shares kept dropping, but analysts who covered the company kept telling clients to buy the stock.[18] Their calculations told them the likely cost of the spill could not possibly come close to the $50 billion drop in BP's market value at the end of May. When one calculated tax impacts and expected contributions towards the bill from BP's partners Anadarko and MOEX, this share drop implied a gross cost of around $100 billion. But the analysts – most of them based in London – were missing the point, and clients who listened to them would lose their shirts. Like BP, the analysts had totally misread the politics of the situation. It wasn't simply a case of making good on the damage. This was America, not Britain, and someone had to take the blame. The US public demanded it, and they demanded it quickly.

The effects of this transatlantic misreading were visible in BP's falling share price. Time and again, Obama or some other government official would make a comment in the evening, Washington time, only for it to be ignored by the UK market when trading opened the following morning. Then, when the US markets opened, BP's New York-listed shares would tank, dragging the London-listed shares down with them during afternoon trade.

Hayward, Gowers and Parker were all British and had never worked for any protracted period of time in America. They simply did not understand that being the biggest company in Britain didn't mean the US body politic would not drive you out of business to quench the public thirst for blood. The Federal Reserve certainly got the picture: even as BP denied it was considering bankruptcy, the Fed began to study the systemic risks associated with just such an eventuality.

By the end of May, the situation facing BP was parlous. It needed some respite, even if only for a little while. The answer it devised was 'Top Kill'.

'I want my life back'

After BP's early – failed – efforts to capture the oil spewing into the Gulf, the company decided on a new plan, which involved pumping drilling mud into the well. The idea was that if they could get enough mud into Macondo, the weight of it would be sufficient to keep the hydrocarbons in the reservoir at bay.

The only permanent way to plug the well was to cement the bottom of it via the relief well. This process had been started in April but would take until July or August to finish. The idea behind Top Kill was that, after stabilising the well with drilling mud, BP could stick in a cement plug. This would seal Macondo until the relief well was finished.

Hayward told the world that Top Kill had a two-out-of-three chance of success.[19] The origin of this statistic is unclear. Even the boss of the well control specialist BP hired to help cap the well, Wild Well Control Inc., has said that, at the time, he ascribed a minimal chance of success to the measure.[20] Nonetheless, BP whipped up a media frenzy around Top Kill. Hayward, Suttles and other executives did a series of interviews talking up the chances of success. I was in London when Top Kill was first mooted and Gowers advised I quickly get to Houston to make sure I didn't miss the excitement.

As it happened, Top Kill was delayed while the US government and BP fretted over whether the additional pressure put on the well might cause it to rupture below the ground. This is what is known as a subsurface blowout, and would result in possibly a hundred small leaks emerging across square miles of seabed: a disaster that no one, even with BP's limitless budget, could hope to contain.

The delay only served to heighten the tension. Across America, and the world, people watched news channels and trawled Internet news sites for confirmation that the measure, which they had been told should end the spill, had started. Mobile communications added to the sense of drama as commuters on public transport networks around the world watched video footage of the plume on iPads and mobile phones, trying to spot changes in its consistency that might indicate Top Kill had started.

I found myself sitting in the Reuters Houston bureau, waiting for an update. Finally, on 26 May, almost a week after the original start date allotted for Top Kill, I was summoned to the Westlake Campus. My colleagues and I assumed it was for a press conference announcing the start of Top Kill. But the BP spokesman insisted there was no press conference and no decision had yet been made on whether to proceed. No one else was invited, only Reuters. The idea was that I would be there so that, if the decision was made to proceed, Hayward would give a briefing on the spot, and I would pool my report to other news organisations.

I jumped into my compact hire car and sped towards BP's Westlake Campus. Down in the lobby, I corrected the receptionist about my accent. No, I wasn't English, I said, I was Irish. The growing anti-British sentiment was prompting me to correct such errors whenever they occurred. The press officer came down to greet me and we took the lift up to the fourth floor. I was led to a small room with wafer-thin walls. I could work here on my laptop, I was told, until I was called down to the crisis centre one floor below to be briefed by Hayward. That is, if the Top Kill operation proceeded. Lest I decided to try to creep down unattended, a six-foot-two, 300-pound security guard was stationed, with a walkie-talkie, outside the door. When I needed to go to the bathroom, he would insist upon accompanying me. A sign saying 'Journalist at work' was placed on the door, in case any loose-tongued employee inadvertently stepped inside.

A couple of hours later, I was taken downstairs, stopping outside the lift to gain yet another security pass, before walking through glass doors into the crisis centre. A larger centre in Houma, Louisiana was co-ordinating the surface response to the oil spill. That involved trying to contain, capture or burn the oil floating on the sea, and cleaning beaches when this effort failed. But they'd be doing that for ever unless Houston could shut off the well. If there was going to be a solution to this crisis, it would be found here on the third floor.

The Houston crisis centre was a permanent fixture built mainly to deal with hurricanes. Every hurricane season it kicked into action to co-ordinate evacuations of oil platforms and manage any repairs to facilities. The extent of the disaster in the Gulf of Mexico was beyond the scale of a normal hurricane and so the centre had to be expanded, taking over adjacent areas normally used for training staff. In keeping with the communications policy of avoiding all language that might suggest calamity, the centre had been renamed the 'Incident Command Center', although no one had thought to take down the signs on the walls that said 'Crisis Center'.

I was led down a long artificially lit corridor lined with photographs of oil installations. Here and there, a poster taped to the wall advertised 'BP Care', a helpline for workers to help them deal with personal difficulties including stress. The corridor widened and I saw 40 or 50 of the best subsea engineers in the world standing outside a converted training room that had recently been christened the 'intervention room'. Here, engineers had worked on ways to actively stop the well from flowing. Next door was an identical space called the 'containment room', where workers dreamt up ways to capture the oil as it flowed from the well and before it was dispersed in the sea.

Both rooms measured about 30 feet by 30 feet and had rows of white laminated tables and were crammed with laptops powered by cables wrapped in yellow 'warning' tape that snaked from the ceiling. The walls of both were covered with maps of the Gulf, diagrams of the equipment on the seabed and projector screens.

Inside the 'intervention room', Hayward, Andy Inglis, Bob Dudley and US Energy Secretary and Nobel Prize-winner Steven Chu were poring over yards of charts and long tables of data. They were engaged in complicated work but had to answer a simple question: was it safe to proceed with Top Kill?

The engineers waiting outside for an answer to this question were dressed in golf shirts, button-down shirts, jeans and chinos. No one wore a tie. In addition to Texan drawls, one heard English, Scottish and Norwegian lilts: the accents of the offshore oil industry. Many of the men – and, with the exception of support staff, they were all men – had come at short notice, packing small bags of clothes that by this stage had been well worn and often recycled by the laundries of the hotels in which they were billeted.

Even Dudley, whom countless BP executives over the years have described to me as 'the calmest man I ever met', was unable to downplay the atmosphere. He came out at one point during the deliberations on Top Kill. 'It's intense,' he told me with a sigh. 'It's kind of like NASA and the Apollo 13 mission in there.' The astronauts on Apollo 13 had only made it back to earth thanks to a series of patchwork fixes they made with the guidance of technicians in NASA's space centre 30 miles south-west of BP's campus. Forty years on and the famous phrase attributed to those astronauts – 'Houston, we have a problem' – was yet again proving an irresistible template for headline writers around the world.

Dudley wasn't the type of man normally given to hyperbole. Until now, he was best known as the former head of TNK-BP, who had been forced to flee Russia by BP's oligarch partners. In 2008, the billionaires had wanted to wrest control of the venture from BP and figured they needed to get Dudley out. They had him followed, had him harassed by TNK-BP's own guards, and, BP argued, had their friends in government stifle his attempts to secure a visa. Dudley ended up hiding out in Paris for months. He had kept his location secret, not so much out of fear for his safety as for the knowledge that, if the Russians found him, they would slap a bunch of crippling lawsuits on him. Despite the obvious drama, Dudley used to recount the experience with about as much passion as a roughneck discussing the works of Doris Lessing. But today he was showing a rare flicker of emotion. 'It's pretty dramatic,' he conceded.

At a quarter to two, the great minds in the 'intervention room' all agreed that the well could take the pressure. Hayward gave the nod and one of the engineers picked up the phone and told the men on the rig to start the procedure. There was no white smoke. Nor any cheer. When word was relayed to the men waiting outside the room, they silently and promptly returned to their posts. They were no less tense than before, but a little more optimistic.

A few hours later, I was led further along the narrow hallway to another room, where I was supposed to meet Hayward. Halfway down, Andy Inglis stopped us. He appeared even less welcoming than usual today. As the boss of exploration and production, he oversaw all of BP's drilling, and this made *Deepwater Horizon* his personal problem. His job – and, if some of those who were calling for criminal charges to be brought against BP managers had their way, his freedom – was on the line.[21] He said Hayward wouldn't be speaking to the press.

There was more toing and froing but finally I was taken to the Simultaneous Operations, or 'Simops', room, where I was supposed to meet Hayward. This was a kind of air traffic control room for the surface of the sea 5,000 feet above the blown-out well. A dozen cheap desks were arranged in a rectangle in the centre of the room. A similar number of men sat at them, in front of computer screens, co-ordinating the movements of a dozen rigs and ships jostling for space in a kilometre-wide circle at the well site. The captain of one of the vessels compared their work to directing a ballet. A large projector screen showed a live diagram of the area being monitored and the vessels moving within it. Five large flat-screen televisions stood on steel pedestals, showing video images of developments on the seabed. One screen showed the blowout preventer, the piece of equipment that was supposed to have shut off the flow of oil and gas before the explosion.

Hayward came in wearing his signature open-necked shirt. (One men's style magazine had even called BP to ask the name of his shirtmaker.) It was almost a month since I had last seen him at the Reuters offices in London, and he looked like he hadn't slept since then. More tanned, certainly, but his famously fresh face was beginning to look a little past its sell-by date. His thick head of dark brown hair showed a few flecks of grey that I had not previously noticed, and had a ruffled look that made it look thinner. Since I had first met him six years earlier, I had always thought he looked about a decade younger that his actual age. Now he looked his full 53 years. And then some.

Hayward was flanked by Gowers and Parker. A cameraman from a local TV station was also there to record Hayward's comments. He asked me who the guy that looked like Oliver Stone was, nodding in Parker's direction. Brunswick's role was not widely known, so the firm wasn't taking much flak for what was increasingly being seen, even in the UK, as a botched PR job. Gowers' high personal profile in the UK prompted a barrage of invective from some in the British media, who no doubt assumed he had more influence with Hayward than he did.

Hayward began to explain progress on the Top Kill operation, speaking in a controlled and deliberate manner that gave the impression that the previous hour had been spent having this spiel drilled into him by Gowers and Parker. 'The operation is proceeding as we planned it,' he said. 'There's a great team of people doing extraordinary things to plug this leak as quickly as we can.' When I pressed him, he repeated his earlier optimistic forecasts about Top Kill: yes, he still believed it had an almost 70 per cent chance of success, and we would know whether it had worked or not within 24 hours.

Twenty-four hours later, BP was still pumping drilling mud into the well.

I got another call from Gowers. Would I like to go with Hayward and an ABC TV news crew to the drill site the following day? It wasn't the kind of offer one turned down – but, I asked, wouldn't the operation be over by then? No, Gowers said, it had another 24 hours to go, but all the signs indicated it was still going well. Naturally I accepted and set off on a six-hour drive through the night to the Houma heliport.

Gowers himself did not join me, precluded by the fact that he had a beard. Safety regulations meant anyone going offshore had to be clean-shaven because facial hair could, theoretically, interfere with the seal of the breathing apparatus one might be forced to don in the event of a fire or crash. The ABC cameraman shaved his off so that he could make the trip, but for Gowers it must have been a bridge too far. He and Parker instead sent along Brunswick's US representative Michele Davis, the former Treasury official.

In retrospect this was a mistake. Hayward, for whatever reason, largely ignored her throughout the day. Had she been closer to him on the helicopter and on the ship, she might have been able to give him some advice that may have saved his career.

When I arrived at the heliport, a large woman with a wide smile asked me if I had met Hayward. 'He's sooo nice,' she told me. 'And cute.' This was the funny thing about Hayward: he was probably the most hated man in America at the time, yet many people who met him liked him. It was a combination of his down-to-earth manner, deliberate attempts to be friendly to low-level employees, and a strangely endearing awkwardness.

It was not long until the white BP Falcon corporate jet touched down and Hayward was driven the short journey to the heliport in a black Chevrolet Suburban with blacked-out windows. The car had inadvertently become his vehicle of choice after a visit to the Texas City refinery a few years previously. Rather than be chauffeured in a large ostentatious limo, Hayward had asked for an SUV. Word got around BP America that Hayward only drove in Suburbans, and now he was met by one wherever he went.

I was sitting on a bench outside the grey one-storey building that was Houma heliport when Hayward walked up the path, wearing blue overalls with a BP logo, steel-capped boots and Aviator sunglasses. I expected to see a bodyguard but there was none. Bob Malone, the former BP America president, had always insisted Hayward travel around the US with security, but Hayward didn't like it and had ditched the practice after Malone left. Even now, as America's least popular executive, he was loath to be chaperoned.

Hayward had spent the night in the control room, monitoring progress on Top Kill, before doing a round of television interviews. He had had no sleep.

The 16-seater chopper flew us over the lush green Mississippi Delta towards the blue sea. Not far from shore, we began to see long strips of brown scum and areas of sheen. Hayward gazed out. 'There's a lot of clear water,' he said, as though trying to convince himself. 'You can see some silver sheen here but most of the area is surprisingly clear in this location.' Even at this late stage, he had not got the message about downplaying the damage.

Our helicopter landed on the *Discoverer Enterprise*, a drilling ship whose grey hull was longer than two football fields. It sat 5,000 feet directly above the leaking oil well. In the sky, the Gulf of Mexico sun beat down mercilessly and beads of perspiration dripped off the rim of Hayward's white plastic helmet.

He said he wanted to stop the leak as soon as possible, because he cared about the environmental damage, because he cared about the livelihoods of those affected, and because it was also in BP's interests to end the crisis as soon as possible. 'I want to stop this thing as soon as anyone,' he told me, as we stood on the helideck, before adding: 'I want my life back.' The comment seemed a little strange, and I noted it for the feature article I was working on. But I did not find it as horrifying as many Americans would when they heard it. Yes, it was a self-centred comment – but deliberately, not inadvertently, so. The point he appeared to be making was that, in addition to all the good reasons he had for solving the problem, he also had selfish reasons to do the right thing. He wasn't asking the people of the Gulf Coast to feel sorry for him. He was trying to show them that his interests were perfectly aligned with theirs. So convinced was Hayward that Gulf Coast residents would be reassured by this perspective that he repeated it again on the chopper back to Houma, Louisiana, adding: 'I don't have a life right now.' He clearly had no sense that the comment suggested he felt sorry for himself and a bit hard done by. Since Brunswick's Davis was not close by, she probably did not hear him say it. Presumably, if she had done, she would have pointed out the inadvisability of expressing such a self-referential sentiment.

A day later, as I was driving across one of the long elevated concrete highways above the swamplands of southern Louisiana, on my way back to Houston, I heard on a local radio station that Top Kill had been halted. It had not succeeded. If I felt crushed, I could only imagine how the people involved felt.

Suddenly, BP was in it for the long haul. The company proposed other interim measures to stem the flow but by now it had lost all credibility. It looked as if everyone would have to wait for the relief well to be completed before the flow could he halted. This would take another two months.

The US public understandably thought the crisis had already dragged on for too long. Now it was clear that the disaster was only just beginning. The response to this realisation was the kind of national rage one might associate with the outbreak of war. A day after Top Kill was declared a failure, amid this powder keg atmosphere, Hayward made the mistake of repeating the comments he had made to me two days earlier, but this time on national television: 'I'd like my life back,' he said.

It is unlikely that, even if the statement had been broadcast with all the qualifications he offered on the drill ship and chopper, Americans would have seen the comments as other than self-pitying. Without such qualifications, the comments appeared deliberately inflammatory. Americans were enraged. Hate mail was sent to Hayward's home and family members received threatening phone calls.[22] The UK police began to provide round-the-clock protection for his family. The political pressure on BP also shifted up a gear. President Obama, previously reluctant to use inflammatory language, took a leaf out of Ken Salazar's book. He told Americans he had assigned people 'specifically to ride herd on BP',[23] and added that he would already have sacked Hayward if the CEO worked for him.

Gowers had mentioned to me a day after the failure of Top Kill that he was about to go to Washington to meet his new head of US press, Dick Cheney's former press officer Anne Kolton. I reported news of the hire, and the story prompted considerable press commentary in the US, especially among left-of-centre commentators who saw the appointment as further evidence that right-wing politicians and Big Oil were all in bed together.[24] As it happened, Gowers' trip would be for nothing: although no stranger to hostile press coverage, Kolton found the anger towards BP too hot to handle and backed out of the job. A month earlier Hayward had told me he thought Gowers was doing a 'great' job, a comment I interpreted as meaning that Gowers had not asked him to do anything he didn't want to do. Now, amid the media barrage, communications between Gowers and Hayward broke down. It would take a phone call from Gowers' predecessor, Roddy Kennedy, to reopen them.

Elsewhere, BP's scattergun approach to media management created more problems. The company bought terms such as 'oil spill' from search engine providers including Google to help direct Internet users to its website, rather than to news or protest sites. When news organisations (some of whom had also sold BP rights to the term on their own websites' search engines) reported this, it made the company seem controlling and dishonest.

Even worse damage was done when it was discovered BP had doctored images of the relief effort that were made available to the media. BP published one photograph of a helicopter that appeared to be flying near the spill site, with an armada of response vessels visible through the chopper's windshield. The original picture was in fact of the chopper sitting on a helideck of one of the vessels. The helideck had been removed and the vessels visible through the windshield were made larger and clearer, although eagle-eyed critics immediately spotted a slapdash Photoshop job. Another photograph, of workers at the Houston Crisis Centre staring at banks of video screens showing footage of subsea work, had been amended so that some of the screens, which had originally been blank, were now filled in with images of subsea equipment.

The alterations were relatively innocuous and can hardly have been intended to materially improve the perception of the spill. But when the story circulated, people assumed that BP would also be airbrushing the oil out of photographs. BP blamed an 'overenthusiastic' photographer, although clearly poor oversight was the main problem.

All these gaffes increased pressure on the company. The shares continued to plummet, wiping over $100 billion off the company's value at one point. Fears grew that BP's oil leases in the Gulf of Mexico would be revoked, and people openly talked about the possibility of collapse.

In a desperate move, the company splashed out $50 million on prime-time advertising slots in which Hayward apologised to America. Just a few weeks earlier, Gowers had told the *New York Times*: 'In our view, the big glossy expressions of regret don't have a lot of credibility.' BP was clearly grasping at straws.

The campaign enraged Obama, who noted that the amount it cost was about the same as the amount BP had spent compensating victims in the Gulf area. 'What I don't want to hear is, when they're spending that kind of money on their shareholders and spending that kind of money on TV advertising, that they're nickel-and-diming fishermen or small businesses here in the Gulf who are having a hard time,' he said.

BP's board of directors had been watching from the sidelines as the media mess unfolded. By early June, they decided they could stand by no more. It was clearly time for a fundamental change in strategy, and a new spokesman.

7

Capitol Punishment

In American politics, as in the American media, the *Deepwater Horizon* disaster was slow to creep to the top of the agenda. Massive flooding across the Midwest meant that the government already had its hands full on the disaster front. In the week before the explosion, President Obama had signed Disaster Declarations for two states; on the day after it, he signed another two. Washington also had its hands full on the corporate demon-slaying front. A Congressional inquisition of Goldman Sachs executives was days away, and the usual pre-appearance programme of grandstanding by committee members, and carefully timed leaks of stories embarrassing to the invited party, had already begun. The American economy was still staggering after the credit crunch, and the Goldman hearings were an opportunity for Washington to be seen to give a cathartic kicking to an organisation seen as one of the main culprits.

Even outside of this context, an explosion on an oil rig wasn't something that would normally require presidential attention. But an oil spill was something else entirely. Hence, on the day *Deepwater Horizon* sank and made a spill look more likely, the White House stepped up a gear.

President Obama convened a meeting of the heads of agencies and departments that would be involved in any spill response. Photographs of him sitting in the Oval Office surrounded by Thad Allen, head of the Coast Guard, Interior Secretary Ken Salazar, Homeland Security Secretary Janet Napolitano and other officials were circulated to the media to emphasise the sense of action.

A slick had formed on the water as the rig sank but the Coast Guard believed this was caused by fuel in the platform's tanks rather than by a leak from below. Consequently Obama felt secure enough to take a planned weekend break with his wife to Asheville, North Carolina. He flew in on the Friday, three days after the explosion, stopping off for a lunch of barbecued ribs at a laid-back local BBQ joint, before checking into the luxury Grove Inn Resort, a traditional presidential favourite. Over the next day or so, he went hiking in the Blue Ridge Mountains, played two rounds of golf, and enjoyed more of Asheville's famed farm-to-table dining. But by the time he sat down to dinner on the Saturday night, the leak had been discovered. He returned to Washington the next morning.

It was bad timing for the president. Three weeks earlier, at Andrews Air Force Base, he had stood in front of an F/A 18 fighter jet converted to run on biofuel and unveiled a plan to expand offshore oil drilling. He challenged those in his party who argued that offshore drilling was not safe, saying oil rigs didn't really cause spills these days. The move was an attempt to paint himself as a centrist, but also to prod Republicans into supporting a planned energy and climate change bill that would include limits on carbon dioxide emissions, an issue he had outlined as a priority at the outset of his presidency. But Republicans opposed CO_2 caps, believing they would make the US economy uncompetitive against China and other countries with no such restrictions. Offshore drilling and an expansion of nuclear energy were the sweeteners to tempt them to back CO_2 caps.

For over a year, Senators and Congressmen had worked to sculpt a compromise energy and climate change bill. Democratic Congressmen

Ed Markey and Henry Waxman had squeezed a bill through the lower house, the House of Representatives, which included a 'cap-and-trade' mechanism similar to Europe's Emissions Trading System. Democratic Senator John Kerry had then taken up the ball in the Senate. But his party could not go it alone. He needed Republican votes to get a bill through. Kerry teamed up with Republican Senator Lindsey Graham, who in principle was happy to concede to CO_2 emissions restrictions in return for more domestic oil drilling and permits for nuclear power stations. Independent Senator Joe Lieberman also joined the effort.

But Republicans, and even some Democrats from states who relied heavily on the coal industry, which would be penalised by CO_2 restrictions, proved resistant. Indeed, the very term 'cap-and-trade' was anathema to many Republicans. Conscious of this, Kerry, Graham and Lieberman deliberately avoided using the phrase in relation to their proposed bill. As April arrived, time was running out for a deal. The mid-term Congressional elections were scheduled for November, and legislators' focus would soon shift from lawmaking to electioneering. In the event, Obama's olive branch on offshore drilling failed to have much impact in swaying those opposed to CO_2 caps. Progress on the climate change bill slowed.

Meanwhile, oil prices were rising. Since Obama's inauguration, the collapse in crude prices had gone into reverse, taking US gasoline prices from under \$2 a gallon to almost \$3 a gallon by late April 2010. This had the same impact it always did: increased public support for oil drilling. A poll in early April showed that 72 per cent of Americans favoured more offshore drilling. This bolstered the Republicans' favoured policy option: an 'all-of-the-above' energy plan, which, despite the inclusive title, omitted one key part of the Kerry–Graham–Lieberman plan: CO_2 caps. Indeed, the only sop it gave in return for more offshore drilling and nuclear power was minor additional subsidies for renewable fuels. For Democratic representatives such as Ed Markey and Henry Waxman, who had spent over a decade trying to push though legislation to limit CO_2

emissions, and who a year earlier had felt so close to victory, the 'all-of-the-above' plan represented a total failure. They vehemently opposed nuclear power and the Republican-backed agenda, memorably summed up by Sarah Palin as 'Drill, baby, drill.'

The lack of progress on the climate change bill in Congress led the Obama administration to waver. On 20 April, Kerry, Graham and Lieberman attended a meeting at the White House with Obama's Chief of Staff, Rahm Emanuel, and presidential political adviser David Axelrod. As they discussed the state of play, Axelrod betrayed the White House's lack of commitment. Speaking about Democrats in Congress, he said, 'The horse has been ridden hard this year and just wants to go back to the barn.'[1] It appeared that Obama would be prepared to allow expanded drilling without the quid pro quo of CO_2 caps.

By the time *Deepwater Horizon* had sunk to the bottom of the Gulf of Mexico, it seemed that 'drill, baby, drill' had trumped 'cap-and-trade' on Capitol Hill. From a legislative perspective, it looked as if it was going to be an 'all-of-the-above' bill or nothing. And with gas prices heading towards $3 a gallon, Congressmen knew doing nothing would be a hard sell on the campaign trail.

For the oil companies, this represented an explosive backdrop in which to have an oil spill, and BP found itself at the centre of a highly charged political game. Unfortunately for BP, it was a game whose rules Tony Hayward did not appear to entirely understand.

The rules of the game

Immediately after the spill was announced, anti-drilling Democrats set to work. Congressman Henry Waxman, co-author of the energy and climate change bill that had passed the House of Representatives a year earlier and chairman of the Committee on Energy and Commerce, spoke to Bart Stupak, another Democratic committee member who also opposed expanded drilling. The two penned a letter to BP saying the committee would be probing the disaster, and asking for information. The fact that the *Wall Street Journal* received a copy of the letter before BP did was an indication of how this game was going to be played out.

BP was initially dismissive, privately saying the Congressmen were simply trying to use the disaster to raise their profiles ahead of the elections. This view only betrayed BP's ignorance. Stupak, who was the unusual combination of conservative Catholic and Democrat, had already announced his intention to stand down at the November elections. Waxman, on the other hand, always romped home safely in his Californian constituency. He was famous for his lack of interest in courting voters. 'He does not campaign, makes no TV ads, doesn't so much as put up a yard sign,' the *LA Times* had remarked in a profile.

A day later, Congressman Ed Markey, the other co-author of the Waxman–Markey energy and climate change bill, joined in. This triumvirate issued statements, sometimes several per day, criticising BP and its response to the spill. They forced BP to show images of the oil plume on the Internet. The House Energy and Commerce Committee, on which they sat, grilled BP executives and dug up testimonies from rig survivors, as well as internal emails that undermined the company's attempt to blame Transocean. Half a dozen other Congressional Committees subsequently joined the fray and planned their own investigations. Even Stupak dismissed some of these as 'drive-by hearings' whose aim was purely to win publicity rather than to gather information that could be used in drafting legislation.

Democrats on Capitol Hill knew about BP's woeful safety record and didn't much like the company. But this wasn't really personal. If Macondo had been an Exxon well, their reaction would have been the same. The politicians wanted to make it clear to the American public that offshore drilling, as it was at the time being conducted, was not safe and should not be expanded. Indeed, they wanted regulation tightened. They also knew that, as soon as the leak was plugged, the issue would fall from public view. This meant they had to hit fast and hit hard. The result was a public evisceration of BP that would push it to the brink.

'Boot on the neck of BP'

BP's failure to find a quick solution to shut off or contain the well led to widespread anger among Americans. There was sheer disbelief that an offshore oil industry, which generated tens of billions of dollars of revenue each year, had been developed on the premise that nothing would ever go wrong. Now something had gone wrong and no one knew what to do about it. The American public considered a three-month wait for a relief well to be drilled a less than ideal solution.

Suddenly it was clear that the oil industry had been doing high-rope somersaults without a safety net. Americans were indignant at this, but they also perceived failure within the government: how had the companies been allowed to run such risks? Inevitably, the glare of public scrutiny settled on the regulator, and the image that was illuminated was not a pretty one.

The Minerals Management Service (MMS) had been a problem child for years. A 2008 investigation by the Interior Department's Inspector General found that, between 2002 and 2006, employees at the Denver MMS office had been entertained by and had received gifts from oil companies with whom the employees were conducting official business. In some extreme cases, there was even evidence of sexual relations between representatives of the two sides.[2] Another government probe found that, between 2005 and 2007, MMS staff at a Louisiana office had accepted oil company invitations to skeet-shooting contests, hunting and fishing trips, golf tournaments and Christmas parties. One MMS inspector had conducted inspections of offshore platforms belonging to a company with whom he was, at the same time, in talks about a job. (He was later hired.) One staffer told government investigators that oil company representatives had routinely filled out inspection forms, which MMS inspectors subsequently signed.[3]

Democrats blamed the rot on the Big Oil agenda pushed by the Bush administration. Indeed, President Bush had been open about wanting to promote more domestic energy supply: he had removed the requirement on oil companies in Alaska to do environmental reviews when planning new projects, while at the same time forcing the MMS to conduct additional reviews (assessing the potential energy supply impact) for any proposed new regulations. But it was under President Clinton that the MMS had inadvertently made its biggest giveaway to the oil industry. In the 1990s, it had omitted a standard clause from oil lease contracts, which meant that if oil prices went up, the government was entitled to a higher share of the revenue. This went on for a number of years and even when one oil company pointed out the error to a senior MMS official, Chris Oynes, no changes were made. When asked by the Congressional Committee that later investigated the matter, Oynes said he had no recollection.[4] At one point, the Government Accountability Office estimated the mistake could cost the taxpayer $60 billion. Later, after the mistake was uncovered and Oynes was criticised in an official investigation,[5] Bush promoted Oynes to MMS head of the Gulf region.[6]

When Obama came into office he appointed Ken Salazar, a Stetson-and-cowboy-boot-wearing former environmental lawyer, as Secretary to the Department of the Interior, the department that oversaw the MMS. Salazar promised to clean up the MMS, telling oil companies they would no longer be 'kings of the world'. Yet he had done little, if anything, to make drilling safer by the time of the BP spill. This was because he focused on only one side of the agency's remit: revenue generation. He had made some progress on getting a bigger share of oil revenues for the taxpayer. He had also abolished a royalty programme that the oil companies were seen to have used to short-change the government,[7] and started to look at new variable royalty systems that might boost the government's take.[8]

But Salazar had not stiffened safety regulations or boosted enforcement of existing rules.[9] Chris Oynes, who had, according to colleagues,[10] overruled environmental concerns in order to push through drilling, kept his job. No one dusted off any of the studies from the previous decade that stressed the growing risks of pushing into ever deeper water. These included a 2000 MMS study warning that a deepwater blowout could lead to a much bigger spill than from shallow-water wells, and that the industry didn't know how to deal with oil deep under the sea. Or the 1999, 2001 and 2004 studies that warned of the limitations of blowout preventers in sealing off blown-out wells.

The problem here was one of structure. Regulators usually sit outside the industry they oversee, but the MMS was an insider. Indeed, raising over $10 billion per year from lease sales and royalties, one could say it was the biggest player in the offshore oil industry. It was less the industry's sheriff than its principal partner. The MMS's clearest incentive was to keep business ticking along in the Gulf. It was due to this inherent conflict that other countries such as the UK and Norway split the licensing and regulatory functions into two totally separate government departments. By pushing the potential conflict of interest all the way to the prime minister, one removed an incentive to sacrifice safety for the sake of revenues. But the oil industry liked the prevailing US structure and opposed when President Clinton proposed breaking up the MMS. A spokesman for Chevron said the industry liked 'one stop shopping'.[11] Indeed, the industry thought the MMS was so in tune with its thinking that two former directors of the agency had been made president of the National Offshore Industries Association, the offshore oil drilling lobby group.

Salazar did not commit additional resources to the MMS to help it execute its oversight duties more carefully. Lack of resources meant the agency did not have the staff to review all the risk assessments and permit applications oil companies were supposed to submit. To help overcome this, the agency routinely granted waivers for the environmental impact studies stipulated under law.[12] When it came to required permit applications, the frequency with which MMS staff did not spot errors or even contradictory statements suggests they only gave the requests a casual glance before rubber-stamping them.

Because of all this, Salazar must have known there was a risk that public anger could sooner shift in his direction. In a quick move to preempt this, the head of the MMS stood down. (Salazar said she had stood down of her own volition, although White House officials told TV and newspaper reporters that she had been sacked.) Oynes took early retirement. Then Salazar turned on BP, becoming the company's harshest critic in government. A week into the spill, he did the rounds of the Sunday current affairs programmes, declaring: 'Our job is basically to keep the boot on the neck of British Petroleum to carry out the responsibilities that they have.'

Hayward, preparing for a meeting with Salazar the following day, knew BP's honeymoon period with the US government was over.

Calm under pressure

President Obama's reaction to the oil spill was markedly more measured than that of his increasingly vitriolic Democratic colleagues and his angry Secretary of the Interior. Indeed, he rebuked Salazar for his 'boot on the neck' comment. 'I would say that we don't need to use language like that,' he said.

Obama is by nature a considered, reflective person. When asked why he took a couple of days to express anger towards Wall Street insurer AIG, which had been bailed out by the taxpayer but then paid big bonuses to executives, he replied: 'Because I like to know what I'm talking about before I speak.' Such calmness under pressure is often seen as defining leadership. But amid the drama of the oil spill, all the usual rules got thrown out the window, and Obama's refusal to lose his cool was interpreted by some as a lack of concern. Even his own supporters began to criticise him publicly: the Ragin' Cajun James Carville demanded Obama tell BP 'I'm your daddy.'

Republicans depicted Obama's unemotional reaction as dithering. They criticised the spill response effort and said the president had been slow to react. A week into the crisis, the right-wing Fox News TV station began comparing the disaster to Hurricane Katrina. George Bush's botched reaction to Katrina, which cost over 1,500 lives along the Gulf Coast, had for many Americans defined that administration as incompetent and uncaring. Indeed, Fox even invited Michael Brown, who was sacked as director of the Federal Emergency Management Agency for his handling of the Hurricane Katrina response, to comment on Obama's handling of the oil spill. Brown said Obama had deliberately delayed responding to the spill so that it would get bad enough to reverse public support for expanded offshore drilling.

Meanwhile, Sarah Palin said Obama had handicapped the oil skimming effort by not waiving the Jones Act, which required ships travelling between US ports to be manned by US workers.[13] She said this Act prevented foreign skimming vessels from entering the country to help with the spill, and that Obama's continued enforcement of it was because his friends the labour unions saw the Act as defending their members' jobs. In fact, this was nonsense. Thad Allen, head of the response effort, went on TV to say no one had yet asked the president to waive the act; if asked, he said, the administration would waive it.

Rudolf Giuliani, the former Republican mayor of New York and hero of 9/11, appeared on TV (MSNBC this time) to criticise Obama's handling of the spill and his decision to take a weekend break at the beginning of the crisis. 'This has been a gross failure in crisis management,' he said. 'Could not have done it worse.'[14]

All of this made Obama appear impotent. It was easy to paint him in this light, even though most criticisms of the response effort, including the Jones Act accusation, did not stand up to scrutiny. Critics could tell him to federalise the response effort but everyone with an understanding of the situation knew this wasn't possible. The only party who could fix the problem was BP. If the government did take over the job, it would only involve putting even less competent people in charge.

But as the crisis dragged on, the pressure on Obama began to mount. Though his aides continually denied that the oil spill was his Hurricane Katrina, polls showed Obama's approval ratings hit the same lows Bush's did in the wake of that disaster. There was really only one direction in which he could redirect public anger: towards BP. He began to ratchet up the pressure on the oil giant, dropping his calm façade and appearing visibly angry on TV. He adopted more robust language, telling BP – or 'British Petroleum', as he took to calling it – to 'plug the damn hole'. In another interview, he explained that the reason he spent so much time talking to experts was so that he knew 'whose ass to kick'.

Fever pitch

Tony Hayward had started out seeing the oil spill as a technical problem. Never having worked in the US he had little understanding of the political landscape,[15] and was perhaps uniquely ill-equipped to deal with the post-Top Kill political onslaught. Indeed, his focus on the brass tacks of BP's business had led him to overlook such activities as glad-handing politicians; former aides of the more worldly John Browne were amazed to discover that, after three years in the job, Hayward had never even met Obama.

Before the oil spill, BP had been used to an altogether cosier relationship with government. While Congress and the White House were seen – especially when Republicans were in charge – as handmaidens of the oil industry, BP had long enjoyed a level of political influence in the UK that none of the US majors could hope for. John Browne had been so close to the Labour government that British newspapers had nicknamed the company 'Blair Petroleum'. His chief adviser, Nick Butler, was a former Labour Party parliamentary candidate who went on to become a special adviser to Prime Minister Gordon

Brown. John Browne had hired Tony Blair's former chief of staff Anji Hunter as BP's head of communications. BP staff sat on government task forces and had secondments with the Foreign Office and the Department of Trade and Industry, while former CEO David Simon had been appointed Blair's Minister for European Trade and Competitiveness. Meanwhile, BP had hired former government ministers and intelligence agency figures as special advisers and company directors.

It was a relationship that helped ensure BP's interests were considered at the highest possible level. When Libyan dictator Muammar Gaddafi agreed to give up his nuclear programme in December 2003 in return for rehabilitation into the international community, BP was one of the biggest beneficiaries. In 2007, Hayward signed BP's biggest ever exploration deal, worth $900 million, with the autocratic government. Six months later, Libyan officials stalled implementation of the deal, telling BP they were unhappy about a delay in implementing a prisoner transfer treaty, which would have allowed the return home of Abdel Basset al-Megrahi, the former Libyan intelligence agent convicted of the 1988 bombing of a Pan Am passenger liner over Lockerbie, Scotland, which killed 270 people.[16] By this stage, Mark Allen, the former British intelligence officer credited with helping negotiate the end of Gaddafi's nuclear programme, was working for BP. He never confirmed that he had once been a spy but he cheekily hung a picture showing a cloak draped across a table with a gem-encrusted dagger on top on the wall of his office on the fourth floor of St James's Square. Allen passed the Libyan comments on to the British government.[17] Soon enough, al-Megrahi was released from prison and on a plane to Tripoli. The reason the Scottish government gave was compassionate release: he had terminal cancer and was allegedly weeks away from death. Following an investigation into the matter in 2010, Prime Minister David Cameron said in a statement to the House of Commons that the previous government had sought to facilitate the release.

This wasn't the only potential barrier to BP's Libyan expansion that the British government sought to remove. The government rejected calls from IRA bomb victims that Britain seek compensation from Gaddafi, on the grounds that he had supplied the explosives that had injured them. The US government had sought and received compensation for the few Americans caught up in these blasts, but Britain demurred, fearing a negative impact on Anglo-Libyan business dealings, the biggest of which, by far, was BP's agreement. When news of the government's decision was revealed, it did a rapid and embarrassing U-turn and asked Libya for compensation.[18]

Finally, the British government also abandoned any attempt to try the murderer of an unarmed policewoman who had been shot outside the Libyan embassy in London. After the killing, this individual and the rest of the Libyan embassy staff had been allowed to leave the UK. Britain had for years insisted that normalisation of relations with Libya required the killer to face justice in the UK, but did not follow through on the policy amid concerns that Britain's commercial interests could be harmed.[19]

Given BP's track record with the British government, it is perhaps understandable that Hayward continued to view the oil spill as a technical problem rather than as the biggest political assault faced by an oil company since 1906, when Teddy Roosevelt had started the fight to break up John D. Rockefeller's Standard Oil. He designated Lamar McKay, president of BP America, to handle relations in Washington while he oversaw events in the Gulf. But US politicians would not be fobbed off with what they saw as a glorified PR man. One Congressman of Japanese extraction was sufficiently unimpressed with McKay to suggest, at a Congressional hearing, that he should commit hara-kiri, ritual suicide with a samurai sword. Some investment analysts were shocked by McKay's failure to back up BP's claims to have fundamentally improved safety since Texas City. 'When asked to specifically identify how,' analysts at investment bank UBS said in a research note, 'BP was unable to do that in a convincing manner.'[20] Of course, they might as well have asked themselves why they believed the company had improved.

By June, the Macondo well had been flowing for over a month and it was open season on BP in America. Robert Reich, Secretary of Labor under Bill Clinton and a leading political commentator, argued[21] that the government should nationalise BP's assets. Ken Salazar said BP should pay for all the jobs lost as a result of the ban on deepwater drilling that Obama had imposed in the wake of the spill, creating a new multibillion-dollar liability. Congress debated a bill that would bar BP from receiving future licences in the Gulf.

BP's credit rating was cut and its credit default swaps – a kind of insurance on its debt – began to trade at junk levels. Investors began to wonder where it would end. The crisis in confidence caused a cash crunch: the Bank of America refused to buy crude from the company;[22] BP's suppliers wanted to be paid up front while the company's lenders were growing unwilling to extend credit. Like many large companies, BP relied heavily on short-term debt markets to fund itself, borrowing up to $10 billion at any one time. But it suddenly found itself with no one willing to buy its debt. 'The capital markets were effectively closed to BP,' Hayward later said.[23] 'We were not able to borrow in the capital markets either short or medium term debt at all.' His successor Bob Dudley later admitted that BP faced 'a corporate crisis that threatened the very existence of our company'.

BP was technically solvent: its assets were worth far more than its liabilities. But this didn't mean it couldn't go out of business.

Cold shoulder to cold shoulder

Across the Atlantic, BP's plight was being watched with increasing sympathy and, indeed, frustration. Britons detected a distinct anti-foreigner bias in the constant reference to 'British Petroleum', and while US officials denied any such bias, comments from senior public figures in America undermined this claim. In a Tweet, Sarah Palin urged Gulf

Coast residents to 'learn from Alaska's lesson w/foreign oil co's: don't naively trust- VERIFY' – apparently ignoring the fact that Alaska's biggest oil spill was caused by Exxon. Democratic Congressman Anthony Weiner declared in a TV interview that, 'Whenever you hear someone with a British accent talking about this on behalf of British Petroleum, they are not telling you the truth. That's the bottom line.'

The British press reacted with predictable indignation. 'This great British company has done its utmost to find a solution to the oil spill disaster in the Gulf of Mexico,' wrote the *Daily Express*. 'Obama, anti-British as ever, has simply sat on the sidelines and carped.'

The notion that Obama was inherently anti-British had previously been sparked by the revelation that Obama's grandfather, a Kenyan, had reportedly been tortured by British colonial forces during the Mau Mau rebellion. Some British newspapers had perceived a snub in Obama's decision to remove a bronze bust of Winston Churchill – a post-9/11 gift from Tony Blair to President Bush – from the Oval Office. Obama had replaced it with a bust of President Lincoln, his political hero, although the motivation may well have been to signal his break with the post-9/11 worldview espoused by his predecessor. But the oil spill prompted some newspapers to take the suspicion of presidential anti-Britishness to paranoid proportions, criticising Obama for not holding a state dinner for the Queen when she visited America during the crisis. It would have been tough to squeeze one in, given her US visit was a five-hour stop-off in New York to address the United Nations.

In their coverage of fraying international relations, British newspapers also cited the Bhopal disaster as a way of countering BP's American detractors. A quarter of a century earlier, an explosion at a badly maintained chemical factory in Bhopal, India, owned by American multinational Union Carbide, had emitted a poisonous cloud that killed thousands of people. In June 2010, in the middle of the BP oil spill, an Indian judge found Indian executives of the company guilty of negligence; Prime Minister Manmohan Singh asked the US to extradite the company's former CEO, Warren Anderson, to face charges, something the US had previously refused to do. British newspaper commentators saw hypocrisy in the US's treatment of BP alongside its indifference to the Bhopal victims: it looked to them as if the Americans didn't mind their countrymen causing industrial disasters overseas but were ruthless when it came to a foreign company having an accident on US soil. There was a point to all this, although Britain itself had displayed hypocrisy on the subject of CEOs at the helm during deadly accidents overseas, but whose reputation remained untarnished at home.

But it wasn't simply the injury to national pride that raised British heckles. BP was the biggest dividend payer in the UK. Millions of pensioners relied on the BP payout for their income. When Obama started questioning whether BP should be paying such big dividends, the UK love affair with the president came to an abrupt end.

BP tried to tap into this growing nationalistic mood. Chairman Carl-Henric Svanberg held meetings with government representatives, and in early June the company hosted a party for the great and the good of British business.[24] Attendees included Sir John Rose, CEO of jet engine maker Rolls-Royce; Marcus Agius, the chairman of Barclays; Sir Martin Sorrell, CEO of advertising giant WPP; Chris Gibson-Smith, chairman of the London Stock Exchange; Frank Chapman, CEO of gas producer BG Group; John Buchanan, chairman of Smith & Nephew and BHP Billiton; Vittorio Colao, head of Vodafone Plc; and Richard Lambert, head of Britain's main business lobby group, the CBI. Also in attendance was Sir John Sawers, head of MI6, the UK's overseas spying agency. MI6 refused to say whether he was there in an official capacity or simply networking in advance of a future private sector move. The *Financial Times* noted that 'there was barely an American accent in the room', and said there was 'moaning about a rampant media stoking public outrage in the US and much anguish about the Obama administration's hostility toward BP'.

The party and backroom networking paid off, and the British business establishment came out batting for BP. 'It's a matter of concern when politicians get heavily involved in business in this way,' the CBI's Lambert declared, as though millions of gallons being spilt in the sea was a purely commercial matter. The Institute of Directors said it was time for politicians to act and London mayor Boris Johnson told the BBC it was 'a matter of national concern if a great British company is being continually beaten up on the international airwaves'.[25]

Despite such support for BP, the British government was slow to answer the company's plea for help. Prime Minister David Cameron's hesitation was partly due to transition – he only moved into 10 Downing Street in mid-May – but he was also desperate for his first visit to the US, scheduled for July, to be a success. He needed it to help counteract his reputation as a neophyte on foreign affairs, and feared a harsh statement on the oil spill could poison relations with Obama. The prime minister's dithering drew harsh criticism from the press, business groups, shareholder groups and political leaders from both the Conservative and Labour parties. Then, when he did finally wade in on the issue, he seemed to side with the Americans rather than BP. 'This is an environmental catastrophe,' he said. 'BP needs to do everything it can to deal with the situation and the UK government stands ready to help. I completely understand the US government's frustration.'[26]

It wasn't the kind of message BP was hoping for. Fortunately for BP, hostile media reaction to this comment led to a more forceful statement, hours later, from Chancellor of the Exchequer George Osborne. 'BP is a strong, viable company and it is in all of our interests that it remain so,' he said, in effect chastising the US government for its attacks, which put the company's viability at risk. At last, it seemed that someone was going to stand up for BP. Cameron repeated the message to Obama in a telephone conversation a few days later.

The company was still at its lowest ebb. Its shares were half their pre-blast level. Yet the sheer desperation of the situation shortly gave way to a solution.

Shake-down

On 14 June, Tony Hayward and executive director Bob Dudley flew from Houston to Washington DC. They checked into the Jefferson, a luxury hotel just four blocks from the White House. The men were joined by BP's legal team, who had spent recent days feverishly haggling with White House officials about a plan that could take the heat off BP, and the US government.

The following night, Obama raised the pressure on the company by giving his first ever address to the American people from the Oval Office. The setting was not one chosen lightly. Oval Office addresses are normally reserved for grave crises such as the September 11 attacks and the Cuban Missile Crisis. In the address, Obama spelt out that he wanted BP to set aside cash to pay for the spill. Even though BP had no legal requirement to do this, it was, in principle, happy to make such a sacrifice. But it feared the administration would use such a fund as a piggy bank to pay for any measure they liked, even if these had only a tangential link to the spill.

On the morning of 16 June, Hayward, Dudley, Carl-Henric Svanberg and others from the BP team walked up Executive Avenue, a closed-off street alongside the White House, and passed through a side entrance to the West Wing. Svanberg led the way with a firm face, followed by a worried-looking Hayward. It must have felt like being summoned to the headmaster's office.

The team was led into the Roosevelt Room, a windowless meeting room across the hall from the Oval Office. Here, under the gaze of a portrait of President Teddy Roosevelt astride a prancing horse, they found their names on white cards laid out on one side of a long, highly polished oak table. The cards on the other side of the table carried the names of half the American cabinet, including Ken Salazar, Janet

Napolitano, Attorney General Eric Holder, senior adviser Valerie Jarrett, Chief of Staff Rahm Emanuel and Secretary of Energy Steven Chu. The government members streamed in and the meeting started. Almost an hour later, Obama and Vice President Joe Biden came in. They stayed about 15 minutes, long enough for an official photograph to be taken of Obama admonishing the executives.

By the end of the two-hour window that had been allocated to the meeting, BP had agreed to the White House's demand that it set aside $20 billion to pay compensation claims from people and businesses affected by the spill. The company also agreed that payouts from this fund should be administered by someone independent of BP and approved by the White House. But BP still needed assurances that the fund – which did not cap BP's liability – would not be used to pay for items BP considered not directly related to the spill, such as loss of jobs due to Obama's drilling ban. After another two hours, the two sides had reached an agreement. Contrary to what Salazar had declared days earlier, BP would not be forced to compensate workers or businesses hit by the drilling ban. It was also agreed that BP could finance the fund over four years, rather than be forced to find the cash at once.

Best of all, BP had received what it saw as an implicit agreement from the administration to tone down its rhetoric against the company.[27] The executives believed the administration had got the message that BP's shoulders were only so broad. If it continued its war on the company, BP would be killed. While this would raise a cheer among many voters, it would leave Obama with an even bigger problem: he would be left trying to cap and clean up the leak alone.

Svanberg was invited into the Oval Office for a one-to-one meeting with Obama. The two sat on blue and white upholstered armchairs in front of the fireplace, looking back at the presidential desk. The chairman detected no animosity in the president's demeanour: if anything, he sensed relief in Obama's tone, perhaps at the fact there was someone other than Hayward to deal with at the top of BP. But any relief Obama felt was eclipsed by the sheer sense of liberation Svanberg experienced when Obama declared at the outset that 'I don't believe this couldn't have happened to any other company. I think this is an industry issue.' It was music to his ears. If only BP could convince the rest of the country of this, it would be on the road to recovery.

Svanberg considered the White House meeting a near perfect success for BP, and for himself: 'I probably spent more time in the Oval Office than any other Swede in history,' he boasted to friends. He would later tell investors that he had stood up to Obama. 'He said to me, "You have to pay the fishermen," and I said, "Yes, but we have shareholders who rely on the dividend and we have agreed to cut that and it has caused them a lot of pain."'[28]

The only black spot came on departure, when Svanberg, followed by Hayward doing the strange-looking tongue-pressed-against-cheek gesture he had developed over previous weeks, walked out to a barrage of waiting media. Svanberg declared that BP was trying to do the right thing because it cared about 'the small people'. It was an attempt to sound genuine by speaking colloquially. It failed, and necessitated yet another BP apology.

The White House touted the deal as evidence that the president was making good on his promise to 'make BP pay'. Nonetheless, Republicans were disgusted at the way a government had reached into a private corporation and basically seized $20 billion without legal basis. One Congressman dismissed the deal as a 'shake down' and, though a public outcry saw him quickly retract the statement, right-wing commentators continued to use the term to describe the deal.

For BP, however, the White House pact was everything it wanted. The fund wasn't a financial hit: the company had already said it would pay for the damages that the fund was now being established to cover, so it didn't open up new liabilities. Since BP would channel the $20 billion into the fund over four years, the deal didn't even put additional pressure on the company's cashflow. What BP got in return was a partial U-turn on the liabilities it was being asked to cover, and an end to the daily flogging by government officials, which had been crushing investor and creditor confidence.

Little wonder that BP's shares soared over 8 per cent after the meeting. Those few analysts that had been bearish about the stock suddenly began to issue 'buy' ratings; Bill Gross, boss of Pimco, the world's largest fixed-income fund manager, said he was buying BP bonds. The company was back on track. All it had to do now was 'plug the damn hole'.

8

Subsea Slip-ups

When the leak at the bottom of the Gulf of Mexico was discovered, Tony Hayward put Andy Inglis, head of exploration and production and Hayward's would-be heir, in charge of the subsea response effort. Inglis sent out orders to dozens of senior BP engineers around the world, urging them to come to Houston as quickly as possible. Many packed small suitcases, not realising it would be months before they saw their families again. Some had their arrival delayed by a cloud of volcanic ash that blew in from Iceland and shut down European air space. But one way or another, they all came.

The headquarters of the subsea operation was established at the permanent crisis centre on the third floor of the main Westlake Building. Five to six hundred people worked here in 12-hour shifts, although handover periods meant they all worked at least 13 hours a day. Five meals were served around the clock, usually from large aluminium foil containers laid out at the crisis centre coffee station. Southern cooking was the speciality: barbecued ribs, fried chicken and French fries, with freshly baked chocolate-chip cookies between meals. If staff felt they could get away from their laptops for longer, they might venture up a couple of floors to one of the canteens, where the caterers laid out steel dishes of food and where they could sit down at dining tables. The downside of visiting the canteen, however, was that it entailed having to listen to the TV tuned to CNN that hung from the ceiling. 'So are they accusing us of covering stuff up again?' one woman asked angrily as she walked in on one of the days I visited.

To relieve some of the stress, BP had set up a couple of massage chairs in an alcove on the third floor, where workers could get back and neck rubs. Despite the five meals a day, stress meant some staff were losing weight; a local dentist reported BP staffers coming in with broken teeth due to grinding.

Even with the sum of BP's expertise, Inglis soon realised he needed more help. Within a week, Tony Hayward was on the phone to his opposite numbers at Chevron, Shell and other oil companies, asking for people and materials. BP also called in Wild Well Control Inc., a well-capping specialist that had previously helped seal the Kuwaiti oil wells set alight by Saddam Hussein's forces at the end of the first Gulf War. Wild Well's team was led by Pat Campbell, a famous figure in the oil industry who had started out with the best-known wild well-capper of them all, Red Adair, whose exploits were immortalised by John Wayne in the movie *Hellfighters*. The US government sent down a few staffers from the Minerals Management Service (MMS), and some Coast Guards who strode around Westlake in dark blue jumpsuits and black boots. But in the pre-Top Kill days at least, their presence was largely symbolic.

Photos of the crisis centre later handed out to the media by BP showed a balanced mix of ethnicities and a number of important-looking women, although it was almost exclusively middle-aged white men that I saw during the two days I spent at Westlake. Apart from the tension as everyone waited for Top Kill to start, the mood in the crisis centre was generally calm; nerves undoubtedly taut but not frayed. The workers didn't raise their voices and no one seemed to be dashing around with any sense of panic. Like many high-pressure work environments – newsrooms among them – the atmosphere in reality had little of the drama that movies portray.

The effort to cap the leak was literally without financial constraint. The only limit on what assets or personnel could be employed was global supply. If something was needed, it was purchased from the supplier who could deliver it soonest, irrespective of any additional charges. Even if there was only a small chance something would be needed, it was bought, and then discarded if not used. If the team needed something that simply didn't exist, someone would be hired to design and build it.[1] 'Whatever you needed, you got it,' said one contractor working on the response effort.[2] 'If you needed something from a machine shop and you couldn't jump in line, you bought the machine shop.'

Make it up as you go along

No one had previously considered how to cap a leak in 5,000 feet of water, so the team had to make it up as they went along. Chief operating officer Doug Suttles and executive vice president Kent Wells professed great pride in the speed and inventiveness of the men in the intervention and containment rooms. But oil industry engineers looking in from the outside expressed feelings less of awe than of surprise.

BP's first engineering solution was the cofferdam, a 100-tonne steel containment dome that it tried to lower over the leaking end of the riser, to capture the crude as it escaped. This failed because of a build-up of hydrates – crystallised natural gas – inside the dome as it was lowered in the sea. These hydrates clogged up the pipe that was supposed to evacuate the oil. Andy Inglis, who sustained himself through the long days in the crisis centre with large cups of Starbucks coffee, remarked with disgust to one high-level government official that, 'If we had tried to make a hydrate collection contraption, we couldn't have done a better job.'[3]

In fact, few in the industry were surprised that the dome had not worked, since hydrate formation at such depths is a well-known phenomenon. Indeed, while BP told the Department of the Interior that the probability of success was 'medium/high', Suttles later admitted that BP's engineers had rated the chances of success as 50 per cent at best.

BP next considered trying to place a smaller dome over the riser, in the hope this would not suffer the same hydrates problem, but this idea was dispensed with in favour of sticking a siphon tube inside the riser, to suck oil up to a ship on the surface. Though successful, this only captured a small fraction of the oil being emitted.

The question that outside engineers and academics were asking throughout all this was: why didn't BP simply try to cap the well? When BP finally did just that, three months after the blowout, the well cap was successful. The oil stopped leaking on 15 July. Granted, it was the first time anyone had ever capped a well at 5,000 feet and the equipment used was specially designed for the task; it was not a solution that could have been deployed immediately in late April. But what surprised the oil industry was how long it took for BP to try it. When Exxon's chief executive Rex Tillerson publicly criticised BP's crisis response in September 2010, he put voice to what many in the industry were thinking. 'Most of the tools brought to bear to solve this problem were things we already had and everybody knew,' he said. 'At the end what was done to contain [the well], it's possible it could have been done at the beginning.'

BP dismissed Tillerson's comments as 'Monday morning quarter-backing'.

But Tillerson had a point: capping is indeed the standard industry practice, where possible, for dealing with blown-out wells. Saddam Hussein's forces had set light to around 1,000 wells on their way out of Kuwait, the vast majority of which were killed by installing caps. Subsea wells differ from their land-based counterparts in one vital respect – they are harder to access – but the logic remains the same: cap it if you can.

BP wasn't the first oil company to decide to drill a relief well instead, but this was usually the preserve of companies whose well head was damaged or who lacked the resources to try a cap deep underwater. The Macondo well head was intact, and even the considerable depth was not an insurmountable problem for a company like BP. Still, the subsequent presidential investigation showed that it was not until mid-June – almost two months into the crisis – that BP began planning a capping mechanism 'in earnest'. Why had it ignored standard well-capping practice?

Under pressure

For oil men and industry consultants, BP's apparent deferment of what seemed an obvious solution to the problem became all the more mysterious as spring turned to summer and Macondo showed no signs of letting up. But behind the scenes, BP was experiencing a series of incredibly complex technical dilemmas.

Deepwater Horizon's blowout preventer (BOP) had been successfully activated, with the flick of a switch on the rig, during a fairly routine hiccup a month before the rig blast. On 20 April, after the crew had abandoned ship and the rig had caught fire, however, the only way BP could activate the BOP was with remotely operated vehicles (ROVs) sent 5,000 feet under water – and this too had failed. The BOP had refused to budge. An examination conducted for the Department of the Interior later showed that the drill pipe had buckled within the device, making it impossible for it to snap shut: a clear design flaw that regulators subsequently sought to address.

According to a Transocean employee involved in the immediate response effort, BP had been reluctant to deploy the ROVs, and indeed delayed doing so for around 20 hours. The reason for this was that BP feared shutting in the well would lead to pressure building up inside it, which in turn could cause the well to rupture and crude to seep into the rock and up to the sea via innumerable cracks: a subsurface blowout.

It is true that capping the well would have led to a build-up in pressure. After all, the wall of any pipe faces minimal pressure as fluid passes through it, but covering one end while the liquid continues to flow in naturally leads to an increase in stress. But BP's concern about the well's integrity was still unusual. An oil well is supposed to be designed to handle the pressure of being capped at the top. It is also unlikely ever to have to bear the full weight of the reservoir, since normal drilling practice means multiple barriers should be in place at any one time. Nonetheless, it is best practice for the pressure rating on the steel casings in a well to be sufficient to take the full force of the reservoir.

Macondo was being drilled into a layer of rock laid down during the Miocene era, between 24 and 5 million years ago, and Miocene oil reservoirs in the Gulf of Mexico have an average pressure of 4,500 pounds per square inch (psi). According to BP's calculations, however, the pressure in the Macondo reservoir was 11,850 psi at the time it was tapped.[4] (Over time, as oil is produced and the reservoir depleted, the pressure in a well goes down.) This was more than twice the average pressure for this type of rock, and over 50 times the pressure inside a decent espresso machine. It wasn't a record, though: wells had been drilled into reservoirs with pressures of over 20,000 psi in the Gulf, without incident.

The only difference between building a high-pressure well and a normal one is that you have to take extra care and use special equipment. Casing, for example, has to be bespoke rather than off the shelf, and this is more expensive. And the higher the pressure, the more bespoke and more expensive the equipment has to be. In addition, the casing has to be able to withstand pressure higher than the pressure of the reservoir, since high-pressure wells tend to contain very hot oil, which weakens the pipes it travels through. A previous group of engineers constructing a well that would tap a reservoir with pressure just above 20,000 psi, for instance, had installed casing with a design rating of 27,500 psi. BP's reservoir was 11,850 psi and had high temperature. The pipe it planned to use should have had a pressure rating significantly higher than this. The pipe it did use, according to the Presidential Commission, was designed to withstand 11,140 psi.[5] So BP's fears were justified: capping a well with pressure of almost 12,000 psi using casing designed to take just over 11,000 psi would indeed risk an underground blowout.[6]

But the situation was potentially even worse than that. If the oil was not flowing up the production casing – the evacuation route for oil and gas in most blowouts – BP had an even bigger problem. The alternative route to the surface was up between the production casing and the outer well casing, a space known in the industry as the 'annulus': a dead zone, usually filled with some traces of drilling mud or tiny amounts of gas. A barrier would have prevented the oil rising all the way to the top had BP used a liner and tie-back system, which it had not.

Furthermore, BP had not installed a 'lockdown sleeve', a device that seals the gap between the top of the production casing and the well head. This provided a route for the oil and gas to escape back into the BOP and up the riser. Exxon, Shell and others usually set the lockdown sleeve at the outset of drilling[7] to ensure such an escape route was blocked. BP tended to set it later. Sometimes it never set one at all: 60 per cent of exploration wells are dry holes, in which case BP saved a day's rig time by not having to uninstall the lockdown sleeve; Shell and Exxon incurred this extra cost. BP denied that its decision not to set the lockdown sleeve at the outset had increased the risk, noting that regulations did not require it to, although when it came to drilling the relief well that finally killed Macondo, under the watchful eye of the government, it installed it at the outset.

Yet the peculiarities of BP's well design did not end there. It also chose to insert small valves known as 'rupture discs' in the outer well wall.

These were supposed to prevent ruptures in the steel casing in the event that boiling oil caused any trapped gas or mud in the annulus to heat up and expand. Rupture discs would snap outward if the annular pressure rose above 7,500 psi, allowing a small amount of gas or fluid to escape and thereby bring the pressure back down to safer levels. Conversely, if the pressure outside the outer casing was above 1,600 psi, they would collapse inward. BP's peers mostly used other means to avoid dangerous annular pressure build-ups. These included insulated production casing, which ensured the fluid or gas in the gap did not heat up. They sometimes pumped compressible fluids into the gap, or simply designed their wells without trapped annuluses. Such methods were expensive.

A serious problem with rupture discs was that, if activated, they could offer a path for oil to escape into the rock formation. BP's engineers feared that, if the oil was coming up the annulus rather than up the production casing, and they capped the well, the reservoir pressure would pop the rupture discs outwards, thus causing a subsea blowout. In short, the inclusion of rupture discs in the well design deterred BP from trying the one tested solution the industry had for tackling a blown-out well.[8]

Throughout May, BP had to rebut constant accusations that its oil capture measures were nothing but diversionary tactics designed to create the impression it was doing something to solve the problem, rather than simply being unable or unwilling to cap the well. BP was indeed not rushing to cap the well, but this was because it feared the measure was not safe. In May, the company conducted tests that it believed showed the oil was indeed flowing up the annulus.[9] It was a critical mistake – BP's own investigation later showed the flow had been up the production casing – and meant any notion of capping the well was taken off the table.

But not for ever. By the end of May, the unchecked flowing of the well for a month had decreased reservoir pressure from almost 12,000 psi to

9,000 psi, according to figures given at a press conference by Admiral Thad Allen, the head of the response effort.[10]

This reduction in pressure made another option viable for the first time since the start of the leak: Top Kill. The plan was to pump drilling fluid into the well to create downward pressure of 9,000 psi, so as to push the oil back into the reservoir and then keep it there. This would rebalance the pressure within the well and reduce the strain on its walls. There was a certain amount of risk that the rupture discs might burst during the short period of high pressure needed to carry this out, but even if this happened the result should only be a small amount of drilling mud leaking into the rock, which was not a serious problem.

Some on the BP team, such as Pat Campbell of Wild Well Control, did not believe Top Kill as outlined would work,[11] and instead favoured capping the well and then pumping in mud: a measure known as 'static kill'. BP said it would try this measure if Top Kill failed, but when it did fail BP returned to trying to capture the leaking oil. It was another month and a half before it tried to cap the well. The reason for this change in plan – and further unnecessary prolonging of the spill[12] – came down to another bad decision BP had made, about a day after the rig sank.

Faulty figures

BP's downplaying of the amount of oil spewing into the Gulf of Mexico caused unnecessary damage to its reputation. It also caused massive unnecessary damage to the Gulf. The inaccurate flow rate estimates meant that the company and the US government operated blindfolded from late April until the middle of June, when the panel tasked with providing an independent estimate finally gave a realistic figure. But meanwhile, the erroneous view of the flow rate poisoned almost every attempt to tackle Macondo.[13]

For example, BP said its cofferdam would collect 85 per cent of the oil escaping from the well. But at that stage, the well was leaking 62,000 barrels per day; the *Discoverer Enterprise* to which the steel dome was connected had a maximum capture capacity of 15,000 barrels per day. The cofferdam was replaced by the siphon tube, which was also supposed to divert oil into the *Enterprise.*

When BP returned to oil capture measures, the first of these, the lower marine riser package (LMRP) cap, was initially connected to the *Enterprise.* Suttles said this would capture 'the vast majority of the oil' and the Coast Guard did not disagree. When the *Enterprise* proved inadequate, BP hooked up a line to a nearby platform, the Q4000, to take another 10,000 barrels per day. Hayward said this, too, would capture 'the vast majority of the oil', and again the Coast Guard did not challenge the statement.

Of course, even the combined capacity wasn't sufficient, and BP found itself scrambling for vessels with larger capacities. This wasted weeks and let tens of thousands of barrels of oil flow needlessly into the sea. For almost a month, Thad Allen was content to approve measures that, unbeknownst to him, could at best capture less than a quarter of the total amount of oil leaking.[14] And for another month after that, he approved measures that were also inadequate, if to a lesser extent.

The team in BP's crisis centre and their government masters continually said they were operating on the principle of caution, but caution would have dictated building lots of spare capacity into the measures designed to stem the flow. The fact it wasn't suggests that, though people like Thad Allen later said they knew the 5,000-barrels-per-day estimate was too low,[15] they didn't think it that much of an underestimate.

But perhaps the biggest tragedy caused by the underestimated flow rate was how it led to the abandonment of all efforts to cap the well, unnecessarily prolonging the spill by around another month.

The day after Top Kill was abandoned, BP told the US government that the measure had failed because the company had been unable to raise the downward pressure above 6,000 psi,[16] and that the most likely explanation for this was that mud had been escaping via the rupture discs. The pressure had not been sufficient to activate the discs outwards, but a complex computer model constructed by BP showed that they may have popped inwards during the blowout on 20 April. This, according to BP's theory, created a flow route that made it impossible to achieve the necessary 9,000 psi pressure by pumping in drilling mud. Of course, the gap in the rupture discs also created an exit route for the oil, and BP calculated that capping the well would push large volumes of oil out through this, causing the thing everyone feared most: a subsurface blowout.

This theory was based on a calculation that the amount of mud pushed into the well should have pushed back the amount of oil gushing out. When this failed to occur, it suggested that there was a leak somewhere. But the calculations had a major flaw: only one side of the equation was right. BP's estimate of the flow rate bore no relation to the actual flow rate, so the calculation and the new theory were plain wrong.

If BP had used a better well design it would have had the confidence to cap the well in early May, when capping devices were available. It would not have erroneously concluded the flow route was via the annulus rather than via the production casing. If BP had not underestimated the flow rate, it would have had no reason to think capping the well was so dangerous in early June. Andy Inglis would later describe BP's efforts to tackle the well as a learning process, but Top Kill was clearly a leap backwards. It was only when those involved in the response effort accepted estimates closer to the true flow rate that engineers started to make consistently good decisions, and made real progress towards capping the leak.

Why did all those involved operate as though the low-ball initial estimate given by BP, and later the 5,000-barrels-per-day figure robustly defended by BP, was accurate?

At the outset, BP had an incentive to temper public perception of the flow rate since pollution fines were calculated on the amount of oil spilt. But the fact it operated as though its low estimates were accurate actually ended up hurting the company by prolonging the crisis.

The US government also had an incentive to downplay the flow rate: the worse the disaster, the more pressure heaped on the president. But one would have to assume a high level of conspiracy to conclude that the Coast Guard leadership lied about the flow rate and deliberately allowed oil to spew into the Gulf, simply to protect the White House from embarrassment.

The solution lies, at least in part, in psychology – and in particular in the phenomenon of 'anchoring'. Anchoring involves a person's perception being influenced to an unreasonable extent by a piece of information that they strongly suspect is unreliable, but which carries weight because it was put out early in the discussion. Business negotiators frequently use anchoring to try to achieve the best price in a deal. For example, a potential buyer will throw out a ridiculously low price at the outset of a negotiation. The buyer knows the seller won't accept this price, and the seller knows the buyer doesn't expect to be able to purchase at a price anywhere near the level. Nonetheless, the figure still influences the seller: it 'anchors' the negotiation. It limits the seller's expectations in a way they don't even realise, and drags down the final selling price.

Volumes have been written on the subject of anchoring and studies have shown that even experts are susceptible to the phenomenon. The human tendency to be anchored meant that, after BP had issued its 1,000-barrels-per-day figure in late April, it was hard for people to shift to considerably higher estimates. Two days after the 1,000-barrels-per-day estimate was issued, a government scientist estimated the flow at 5,000–10,000 barrels per day, yet when the Coast Guard issued its revised estimate, Admiral Mary Landry said 'as much as 5,000 barrels' could be flowing. Even as independent scientists offered figures many times this level, the Coast Guard still operated as though the 5,000-barrels-per-day estimate was broadly correct. As a reporter covering the event, I too found it difficult to ascribe credibility to the higher independent estimates – which subsequently turned out to be accurate – simply because of the low level of the original estimates.

While BP did have reason to downplay the flow rate, there is evidence to suggest even it was wrong-footed by the anchoring effect of its own estimate. If BP had realised what the true flow rate was in late May, it would not have bothered with Top Kill and would have gone straight to trying to cap the well. It would also have added more oil capture capacity early on in the process, rather than relying on the *Discoverer Enterprise*. Or, at the very least, it would have had considerable incentive to take these courses of action. A quicker end to the crisis would have eased the political pressure that, by mid-June, was threatening to send BP into bankruptcy. In the end, it seems that no one benefited from the erroneous flow rate estimates.

Surface success

While the Houston crisis centre dealing with the subsea situation generated most of the drama in the response effort, BP had assembled a much larger team to organise the considerably less controversial surface response effort.

In a nondescript industrial building outside Houma, Louisiana, behind a breeze block and glass façade, around 1,000 BP staff, National Guardsmen, Coast Guards, contractors and environmental experts toiled around the clock in 12-hour shifts. It was run by Mike Utsler, who, until the spill, had been overseeing BP's Prudhoe Bay operations; he later replaced Bob Dudley as head of BP's whole Gulf Coast recovery operation. A bouncy 33-year veteran of BP and previously Amoco, he had a mane of wavy hair which he had intended to have cut the day he was ordered to Houma. He vowed to his co-workers that he would not have it cut until the well was capped.[17]

Visitors to the facility, which was normally a training centre for offshore engineers, received an initially typical health and safety briefing – although, this being Louisiana, it ended with its own special warning. 'Don't walk around outside. There's an alligator out there,' one of the guards told me on the day I visited. 'It's not that big – only as big as the table,' he added without any trace of humour as he pointed at a four-foot-long table in front of him.

At the centre of the facility was the main operations centre, a room of about 50 feet by 40 with long rows of desks cluttered with laptop computers. The now-familiar large flat-screens, blown-up photographs, maps and charts covered the walls. A couple of hundred people worked in the room, dressed casually except for the Coast Guards in navy blue overalls, Coast Guard aviators in khaki jumpsuits and National Guardsmen wearing camouflage fatigues and tightly cropped hair.

The Houma centre's goal was to tackle what staff called 'the blob'. It oversaw a fleet of around 125 aircraft, thousands of boats and tens of thousands of people, which between them sprayed dispersants, burnt or skimmed oil on the sea, placed absorbent booms and cleaned up affected areas of the Louisiana shoreline, the worst affected of the Gulf Coast states. Smaller command centres were established in Mississippi, Alabama and Florida to manage shoreline clean-up in those states.

The surface response effort operated under the 'Unified Command' system, which included the Coast Guard, representatives of the State Governors, the Department of the Interior, the Environmental Protection Agency and other government bodies. The structure had been created long before the oil spill and was designed to cope with any number of tragedies, its constituents varying depending on the crisis. It was this structure, with Admiral Thad Allen on top, which allowed the government to rebut accusations it was not in charge of the response effort. But in practice, the surface response was largely a BP affair. BP staff managed the key areas, alongside token deputies from government agencies.

Some politicians, environmental groups and commentators called on the government to federalise the response effort, at least on the surface. Those involved were of the view that the government did not have the competence to run the operation as well as BP, and the fact that the government never did federalise the operation suggests it agreed. Clearly, the White House was more worried about screwing up the job than it was about looking impotent. (Similarly, when Interior Secretary Ken Salazar asked one MMS employee what he would do if the US government took over the subsea effort, the MMS staffer said he would hire BP or another major oil company to run the process. Other oil companies subsequently said they would have refused the invitation.)

Utsler's task was made easier by the fact that, unlike for those running the subsea effort, flow rate didn't really matter. BP was throwing as many aircraft, skimming vessels, shoreline clean-up workers and as much absorbent boom as it could lay its hands on at the problem. Things were going well, all things considered.

Although some politicians in Washington and locally criticised BP for not doing more on the surface, there was a lack of specificity to the accusations. Ultimately, while the National Oil Spill Commission's report into the disaster slammed BP's subsea efforts, criticism of the surface effort was conspicuously absent.

9

Corporate Cannibalism

Carl-Henric Svanberg had not been the first choice to replace Peter Sutherland as chairman of BP when the latter stood down in 2009. Long before his résumé turned up in the headhunter's in tray, BP had tried to court Niall Fitzgerald, the former chief executive of Anglo-Dutch food and cleaning product maker Unilever. Fitzgerald had a solid gold reputation in the City. Like Sutherland, he had been awarded an honorary knighthood – although, as Irish citizens, they were not allowed to use the designation 'Sir' – and sat on the boards of a number of blue-chip companies and hallowed institutions.[1]

But Fitzgerald had rebuffed the approach. In his reflective years, he had become obsessed with ethical business, regularly giving speeches against bribery and corruption. Heading Europe's biggest oil producer was something his principles simply would not allow. 'At this stage of my life I don't feel like making the kind of moral compromises that supping tea with the Saudi royal family would entail,' he told an associate.

Svanberg wasn't even the second choice for the job. When Fitzgerald declined, Paul Skinner, chairman at the time of miner Rio Tinto, was tapped. BP leaked news of the proposed appointment in order to test the water,[2] and investors welcomed the selection: Skinner was a former head of refining at Shell, knew the industry and had a strong reputation. But Skinner unwittingly ruled himself out of the game before his appointment could officially be confirmed, when the heavily indebted Rio decided to bolster its finances by selling a chunk of the business to Chinese state-owned aluminium group Chinalco. Investors went ballistic when they got wind of this, but Skinner approved the deal nonetheless. Although the sale was subsequently abandoned, his reputation had been damaged in the eyes of BP investors.[3]

Others were also considered for the chairmanship but BP's desire to have a candidate who would commit to a ten-year tenure deterred some of the names on the list. So botched had the selection process become that journalists wondered what grandee of British business would possibly deign to accept the job. It was perhaps inevitable, therefore, that the company took its search beyond UK shores. Either way, Svanberg's appointment took the British, and Swedish, business worlds totally by surprise. At the time, he was CEO of telecoms group Ericsson, and had earned great plaudits for turning the ailing company around after the tech boom and bust of the late 1990s. Before that, he had been boss of Sweden's Assa Abloy Group, helping build it into the world's largest lock-maker through a series of acquisitions.

A tall, handsome 58-year-old with a year-round tan and a full head of hair, Svanberg had a high profile in Sweden – the Swedish Richard Branson, according to British newspapers – but he was virtually unknown in the UK. Those who did know him – London-based telecoms analysts – did not regard him with any great fondness, thanks to his controversial handling of a profit warning in 2007. 'They had an analyst day, they went in front of the market and said everything couldn't be better,' one analyst remembered, 'and two weeks later they had a massive profit warning and the stock did [dropped] 35 per cent in a day.'[4] Rumours of his imminent departure from Ericsson had started circulating almost immediately.

Investors had expected that BP's next chairman would either be someone with industry experience, such as Paul Skinner, or someone with a big international profile, especially in the vital US market. Svanberg had no oil industry experience, and while he was well known in Europe and Asia, he had limited experience of the US, since this had been a weak spot for Ericsson. Consequently, the appointment looked a little odd.

Tony Hayward did a round of interviews to promote the news and spoke glowingly of the man who, technically, was now his boss. He was not amused when I asked about the badly communicated Ericsson profit warning. 'I think that's a pretty cheap shot, Tom,' he answered. The British press didn't agree and ran our stories about the blot on the incoming chairman's résumé. Indeed, despite his vociferous defence of Svanberg, colleagues said Hayward was happy to have a chairman who was perceived as a weak candidate. Hayward had only three years left of his planned five-year tenure and speculation was rife that he wouldn't like the idea of a chairman who might lead board challenges against him.

Svanberg took up his new role at the beginning of 2010. The chairmanship was a three-day-a-week job but he did not limit himself to this as he tried to immerse himself in the role. He already had a home in the UK, and even BP's HQ in St James's Square was already familiar territory to him: it had formerly been Ericsson's London base, although Svanberg had sold it soon after taking charge of the telecoms company, to reduce overheads.

On paper, Svanberg and Hayward had much in common. Both had scientific backgrounds – Hayward trained as a geologist, Svanberg as an engineer – and both had led companies through difficult turnarounds. Both liked to watch sport (football for Hayward, ice hockey for

Svanberg); both enjoyed sailing and beer. Yet the two men had very different personalities. Hayward remained shy, disliking 'cocktail party chit chat', as one friend put it, whereas Svanberg was convivial and a natural socialiser.

At a professional level, the pair of them got off to a mixed start. Renewable energy had long been an area of interest for Svanberg and he made it clear in interviews at the time of his appointment that he hoped BP would do more in this field. But when he and Hayward had dinner soon afterwards, Hayward didn't even want to discuss the topic. In general, the CEO struck Svanberg as narrow and uncurious in his thinking – not the kind of blue-sky thinker he was used to in the tech industry. 'Tony's a strange man,' he once remarked to a friend.

When the *Deepwater Horizon* rig exploded, Svanberg had been in the job less than four months. Much of that time had been spent preparing for his first AGM. The chairman is always the master of ceremonies at this event, and though he can divert questions to management, he cannot display total ignorance of the subjects under discussion. The 2010 AGM had proven particularly controversial: a coalition of environmental groups and ethical investors had managed to get a motion on the agenda, which, if passed, would have forced BP to review its investments in Canada's tar sands. It was a sensitive issue that required Svanberg to squarely address the green group's criticisms about the dirty, CO_2-intensive business of squeezing crude from bitumen-drenched soil.

Consequently, by the time of the rig blast, Svanberg still had little practical experience of BP's business. Initially, this seemed to be manageable: the board was briefed on the disaster and Hayward assured them it was an operational matter. Management had it under control. Even after the spill was discovered, Hayward resisted the board's involvement, telling the directors that it was not a matter with which they should unduly concern themselves. The fact that he made a public statement to a similar effect four days after the leak was discovered, telling reporters he didn't expect BP's position in the Gulf to be impacted,[5] suggests that he had failed to grasp the likely significance of the disaster. Indeed, a week after the explosion, he saw nothing inappropriate in reiterating the company's commitment to driving costs lower – the very policy on which the tragedy would soon be blamed.

Trouble at the top

Hayward insisted early on that he should be the public face of BP's response to the oil spill. Given that American corporations tend to merge the roles of CEO and chairman into one, he decided that it could be confusing – and indeed undermining of his own position – for the Swede to conduct political meetings and press interviews.[6] Consequently, while Hayward was meeting US government officials in Washington, Svanberg flew to India with Bob Dudley, head of BP America, for the opening of a solar power plant that BP had built with the Tata Group.[7]

Come the end of the first week of the spill, however, and Svanberg was having second thoughts about leaving Hayward to handle the crisis alone. He had planned to go on a sailing holiday around South East Asia with his partner, Swedish businesswoman Louise Julian, who had travelled to India with him for the opening. His 80-foot yacht, the *Cygnus Montanus II* (a Latinised version of his surname: Swan Mountain), was moored at a marina in Phuket, Thailand. The crew had been ordered to be ready to depart. But should he go?

In the end, Svanberg decided it wasn't the best time to be seen taking a holiday. After one night aboard the yacht with Louise, he cancelled the trip. Instead he returned to London, flew out to Washington the next day, and then travelled on to Houston and one of the crisis centres in Mobile, Alabama. But if he had expected Hayward might be happy to see him, he was mistaken. The CEO was worried that Svanberg's presence could divert attention away from his own PR efforts. He told Svanberg that if he as much as gave a 30-second interview in the US, it would undermine him. When Svanberg asked to meet Admiral Thad Allen, Hayward was reluctant to arrange it.

While in America, however, Svanberg held the first of what would become weekly board meetings. Because they were spread out across the globe, most directors would dial in rather than attend in person. In addition to this, Svanberg spoke to all the non-executive directors every day and chatted with major investors about the spill.

But all this aside, it wasn't entirely clear what a non-executive chairman should do in such a crisis. Strictly speaking, the role of non-executive chairman as defined by governance regulator the Financial Reporting Council boils down to representing investors' interests: making sure management hears investors' voices and provides them with sufficient information. In the corporate world, the primary role of the non-executive chairman is seen as being even simpler: hire the CEO, and then fire him if his actions are becoming injurious to investors' interests. Operationally, non-executive chairmen are expected to stand back and let the executives get on with it. They have been known to take on a prominent public role in times of crisis, but only if the CEO has asked them to, such as when Paul Myners and Roger Carr vociferously argued against hostile takeover bids for Marks & Spencer and Cadbury, respectively. BP also had experience of calling its chairmen into active front-line service. When BP's oligarch partners in TNK-BP had tried to wrest control of the venture from BP in 2008, Hayward had asked Peter Sutherland to deliver a broadside against the Russians. In a speech in Stockholm, Sutherland denounced the Russian billionaires as corporate raiders and criticised the Kremlin for standing idly by. The spat embarrassed Prime Minister Vladimir Putin and helped bring the battle to an end, albeit with Hayward making major concessions. After the spat, Hayward jokingly referred to Sutherland as BP's 'tank'.

In becoming the face of a public battle, the chairman leaves the way open for the CEO to cut a deal with the opponent away from the glare of the media. But such examples were a poor template for Svanberg, since Hayward had insisted he should be the public face of the crisis. So what was Svanberg's role?

Shortly after returning to London from America in early May, Svanberg was visited by Brunswick boss Alan Parker. In the boardroom at St James's Square, Parker told him that, while Hayward wished to remain the face of the crisis in the US, the CEO thought it would be a good idea for Svanberg to start giving interviews in the UK, where critical press coverage was beginning to emerge. To Parker's surprise, Svanberg declined. Friends said he would have been happy to defend the response effort – but he believed that, if he did, he would sooner or later have to defend the way in which Macondo had been designed and drilled. And he wasn't sure he could do that. As an engineer he knew things didn't blow up for no reason, and as a former CEO he knew that disasters shouldn't happen because some low-level employee made a dumb mistake. The disaster was likely to reflect management errors; defending Macondo and the men behind it, including Hayward, might come back to haunt both Svanberg and BP.

Others thought the refusal reflected the fact Svanberg was out of his depth. Some thought he maintained his silence because he feared that speaking out would damage his reputation in Sweden, where he was seen as the poster boy for responsible business.

Irrespective of the reason behind it, Svanberg's refusal to become the face of BP in the UK marked the beginning of a terminal decline in the relationship between the CEO and the chairman.

Bad press

On 12 May, BP flew a group of British newspaper reporters to Houston to show them around the crisis centre, provide access to Tony Hayward and generally show off the Herculean response effort. The resultant coverage was mostly positive, with the possible exception of the left-leaning *Guardian*, which would have been a hard sell even if Hayward hadn't told its reporter that the spill was 'relatively tiny'. But in the days that followed, the news coverage turned distinctly negative in one respect: towards Carl-Henric Svanberg.

On 13 May, one newspaper described the chairman as 'anonymous'. The following day another published an editorial dismissing him as 'BP's invisible man'. An editorial in a third newspaper declared: 'Newcomer to industry leaves others adrift in political slick'. A fourth said the chairman was 'missing in action'. Suddenly, it was open season on Svanberg. It was undeniable that he had kept a low profile for the first four weeks of the crisis. What was surprising was how everyone cottoned on to it at the same time.

Hayward, BP head of press Andrew Gowers and Brunswick denied they had briefed against Svanberg but the chairman, who had continued Sutherland's practice of retaining an independent PR adviser, was unconvinced. After all, the reports roused sympathy for the chief executive at the expense of the chairman, whose primary job, as the saying went, was to hire and fire the chief executive.

A few weeks into the crisis, the position of the chief executive was firmly on the agenda. When Hayward started to downplay the environmental damage of the spill, speculation that he could lose his job intensified. His gaffes and his unravelling claims that Transocean's systems were entirely responsible for the accident prompted some on the board to conclude that they needed to bring him home before he caused too much damage. Public anger at the failure of Top Kill and Hayward's 'I want my life back' comments solidified this view.

But Hayward was resistant. He had committed to staying in America until the leak was sealed, and though he had broken this promise to return to the UK for his birthday celebrations in May, he didn't want to be seen quitting before the job was done. He agreed to come back in early June for meetings with investors, during which he told a conference call with analysts: 'They've thrown a few words at me, but I'm a Brit, so sticks and stones can hurt your bones, but words will never break them.' He insisted on returning to Houston afterwards and was incensed when Svanberg suggested he hand over the spill response to Bob Dudley. Hayward all but stopped speaking to the chairman, and the two men kept their distance when they were summoned to the White House in June.

But Hayward retained his fans on the board, including Sir William Castell, senior independent director and chairman of BP's safety, ethics and environment assurance committee (SEEAC). The committee's mission statement was to ensure that executive management's processes to identify and mitigate operational risks were 'appropriate in design and effective in implementation'. One City diarist questioned whether Castell and the SEEAC had done their jobs well enough,[8] but most of the business press remained surprisingly silent on, or indeed complimentary of, Castell's role: perhaps a reflection of his popularity in the City.[9]

Svanberg won the board over gradually, however. In early June he was so confident of his position that he gave an off-the-record briefing to a number of newspaper reporters telling them that Hayward would be coming home soon rather than staying until the well was capped.[10] Svanberg had also by this stage concluded that Hayward would have to go because of his performance in America – although the board was not yet fully behind him on this. Consequently, in the few public appearances he did make, he voiced support for the CEO, albeit a qualified support. In a *Financial Times* interview he said that Hayward was doing a 'great job', but only in relation to tackling the leak.[11] He did not defend the Macondo well nor endorse the CEO's broader performance. He told a Swedish newspaper that, 'To change CEO now would be a bit like being in the middle of the ocean in a storm and starting to discuss the suitability of the captain.'[12] Of course this begged the question: what about when the storm passed?

Hayward and his advisers saw through such weak professions of support, but Svanberg's profile was on the up. By the time of the White House meeting, President Obama had already announced that he would have sacked Hayward. Indeed, Hayward had been omitted from the original invitation to the White House; on the day itself, it was Svanberg who was called in to be photographed with the president in the Oval Office. It seemed that at least one person in America understood what the primary role of the non-executive chairman was. And Obama wasn't being shy in highlighting how he wanted Svanberg to exercise his powers.

As the pressure on Hayward mounted through June, another round of reports slamming the chairman began to surface in the British press. One went so far as to declare that 'Shareholders want Svanberg to be the fall guy' rather than Hayward. While some BP investors had been telling my colleagues and me that they were unimpressed with the chairman's role, the sentiment expressed in this article was a new one. And what was strange about the story was that it did not actually cite any shareholders. Rather, the sources cited were 'a person close to BP' and a BP 'insider'.

Then reports began to emerge about Svanberg's brief Thailand trip, which had now developed into a 'luxury cruise'. One headline revealed 'Swedish BP chairman whisked married lover on luxury cruise as oil spill chaos erupted', in reference to Louise Julian's pending divorce. The reports added that Svanberg's visit to the US in early May had merely been a drop-in on the way back from his Asian *séjour*. No source was given for the yachting reports.

After spending weeks criticising the US for being anti-British, it now looked as if Britain was engaging in its own form of foreigner-bashing. In the week before the White House meeting and for a fortnight afterwards, the British press was full of articles attacking Svanberg. This might have been entirely understandable, given that his company was causing a massive oil spill, were it not for the fact that Hayward's press was so much more sympathetic. The disparity was especially startling for anyone, like me, who was travelling back and forth from the US, where Hayward was known, following a *New York Post* front-page headline, as 'the most hated – and most clueless – man in America', and where *Newsweek* had remarked that, 'Of all the mysteries of the BP oil spill, perhaps the most baffling is: Why does BP CEO Tony Hayward still have a job?'[13] The attacks on Svanberg were particularly startling since, at worst, he stood accused of being ineffectual during the crisis. Hayward, meanwhile, had been at the helm of the company during the run-up to its greatest disaster.

Forty per cent of BP's assets were based in America, and US politicians were now calling for them to be expropriated and for BP to be prohibited from doing business in the country in future. BP needed to start rebuilding its reputation in America fast. But back in Britain, the business press was suggesting that, given the choice between an effort led by someone with a US approval rating akin to that of Hugo Chávez, or by someone who had been given the stamp of approval by the president, they would take the former. Even Hayward's disappointing appearance in front of the House Energy and Commerce Committee in June attracted sympathy at home. During this hearing, he had used some variant of 'I don't know' 66 times, despite having been given a list of questions in advance.[14] 'Mr Hayward gave the impression that he was the most incompetent CEO in living memory,' remarked David Buick, a commentator beloved of the BBC. 'If that was the real Tony Hayward, then God help BP, but it wasn't.'

Svanberg appeared unmoved by the criticism he was receiving in the British press, and instead focused his efforts on winning BP's 30 biggest investors, senior employees and the other non-executive directors around to his thinking. Gradually, the lunches with senior BP executives and calls to his fellow non-executives started to turn the tide, in his favour. On 18 June, the day after his Congressional grilling, Hayward returned to the UK just in time to hear Svanberg cut the legs from under him on national television. In an interview on Sky News, the chairman announced that Hayward had handed over the day-to-day running of the response effort to Bob Dudley and would, for the rest of the crisis, be spending most of his time outside America. 'It is clear Tony has made remarks that have upset people,' he said. After Svanberg had moved off-camera, the interviewer, Jeff Randall, noted what he considered the most interesting element of the whole conversation: 'He never really came out with a ringing endorsement that, "Tony is my man."'

The comments appeared to take Hayward totally by surprise. BP had already announced that Dudley would take over the response effort once the well was capped, but that was still at least a month away. In a sign of the confusion within BP, spokespeople denied that there had been any change in Hayward's role. Unsurprisingly, the interview did nothing to improve the decaying relations between the CEO and the chairman.

'He's got his life back'

If Svanberg thought distance would limit Hayward's ability to inflame US public opinion, he was wrong.

Along the Gulf Coast, fishermen were unable to go to sea because the oil spill had forced the government to shut vast swathes of the region to fishing. Nonetheless, the day after Svanberg announced Hayward's demotion from the response effort, the CEO decided he would go sailing. He co-owned a 52-foot racing yacht named *Bob* with Sam Laidlaw, the CEO of Centrica, owner of British Gas, and Rob Gray, an investment banker at Deutsche Bank who had advised BP for 25 years. The boat had been sailed to the Isle of Wight for the Round the Island yacht race, one of the biggest racing events in the British sailing calendar. Hayward planned to spend the day on the boat, watching his 17-year-old son compete in the race.

He later defended this decision by saying he had wanted to spend time with his son, whom he said he had not seen in three months. Friends said

Hayward had always gone to great lengths to attend important occasions for his children. But his claim not to have seen his son in such a long time raised critics' eyebrows, given that he had made four trips back to the UK during the crisis. Many other executives involved in the response effort, such as Mike Utsler in Houma, had not been home once since being summoned to the Gulf Coast. He knew he was running the risk of causing another media storm, but he simply covered up in a cap, sunglasses and a high-collared jacket, resulting in international press speculation that he had hoped not to be spotted.

Unsurprisingly, it didn't work. Photographs of him in a sailing boat found their way into the newspapers, with one reporter remarking that, in his attempt to remain incognito, he had ended up dressed like 'a bank robber'.[15]

Across the Atlantic, Gulf Coast residents saw the yachting trip as a two-finger salute. Senator Richard Shelby, an Alabama Republican, said the move was 'the height of arrogance', adding, 'That yacht should be here skimming and cleaning up the oil.' The *New York Post* branded Hayward 'Capt Clueless'. The White House was also unimpressed, although the administration's remarks suggested they were more amused than troubled by Hayward by this stage. 'To quote Tony Hayward,' said Chief of Staff Rahm Emanuel on ABC television, 'he's got his life back, as he would say. And I think we can all conclude that Tony Hayward is not going to have a second career in PR consulting.' It did not sound as if the White House anticipated having to deal with Hayward in his first career for much longer either. When asked about the trip a couple of months later in an interview with the BBC, he said, 'I'm not certain I'd do anything different.'[16]

When Svanberg heard about Hayward's yachting trip, he could scarcely believe it. He himself had been due to attend the wedding of Sweden's Crown Princess Victoria that same day, but had seen no alternative but to decline his invitation. Any sympathy Hayward had garnered in the UK also vanished after his sailing jaunt. Newspapers accused him of 'cocking-a-snoop to the world' and rallied against his 'arrogance'.

Hayward felt the outcry over the PR gaffe was unfair, telling one of his team, 'I've been shafted.' In the weeks that followed, he held a series of upbeat 'town hall' meetings for BP employees, in which he gave every signal that his intention was to lead the company's recovery.[17] He told investors he had no plans to step down.[18]

Boardroom showdown

Hayward's hopes of keeping his job were not without foundation. Exxon CEO Lawrence Rawl had stayed in his job for years after the *Exxon Valdez* ran aground, despite his disastrous media appearances, while even the blast at Bhopal did not cost Warren Anderson his leadership of Union Carbide.

In truth, Hayward was safe for the time being. No one would have been prepared to fill his shoes while the oil was still flowing. The relief well was not expected to kill the well until early August. If BP dropped him before this, and if the killing of the well were then postponed – perhaps due to a problem with the relief well – the new CEO would also be contaminated by the crisis and his ability to rebuild confidence in the company damaged. Hayward must have known this, and he must also have suspected the directors would be reluctant to get rid of him while the oil was soiling Gulf Coast beaches, an equally unedifying situation for any new CEO to have to deal with.

Hayward and his advisers believed it would be autumn before he faced a putsch. By that stage, everything could be different: the media glare would ebb once the well was capped; the damage might turn out not to be as severe as scientists had predicted, the costs not as high as the market expected, and the public perception of the response effort turned to one of admiration. Hayward's reputation might yet recover. But they had not gauged the mood of the board correctly. Svanberg and some of the directors – including Ian Davis, former managing director of McKinsey, and Paul Anderson, former CEO of miner BHP – were among the first to conclude that, given Hayward's exposure to America, he needed to go. By the time the well was capped on 15 July, the rest of the board had agreed.

Svanberg called in Brunswick boss Alan Parker for a meeting, and asked him how he thought Hayward's removal would play out in the media. Parker was blindsided by the question. 'I think it might be best to sound out the board first,' he advised.

Svanberg smiled. 'What if I told you I already had?'

Now Parker knew Hayward was really in trouble. 'Very clever,' he responded, before withdrawing.

Not only had Svanberg lined up the board behind him, he had also flown to the US to agree with Bob Dudley that he was prepared to accept the top job.

In the days after the well was capped, Svanberg told Hayward he should consider his future. Hayward must have been shocked by the rapidity of the move: no one even knew yet for sure whether the cap on the well would hold. But the message was clear: if it did hold, Svanberg wanted him out.

It certainly looked like the end of the road for Hayward, but he had not fought for almost two decades to become CEO to give up the job so easily. Although Svanberg had given the impression that the board was behind him, Hayward knew the matter wasn't quite so clear-cut. For one thing, there had been no board meeting to decide his fate. A stay of execution, at the very least, ought still to be possible. Needing some basis on which to argue his corner, and knowing that Alan Parker's allegiance was to BP and not him personally, Hayward sought advice elsewhere. He called in Matthew Freud, Rupert Murdoch's son-in-law and head of one of the UK's largest PR agencies, Freud Communications, and Philip

Gould, formerly Tony Blair's pollster. He asked the men to arrange surveys in the US to determine whether his position was, as Svanberg believed, untenable.

The results of the polls run by Gould and Freud over the following days showed that Hayward's position was bad. BP might be able to rebuild its reputation in the US with Hayward at the helm, but it would be very difficult. This wasn't the message Hayward wanted to hear. He knew there were some members of the board who wanted him out, and he knew many investors were sceptical that he could lead BP through a turnaround after everything that had happened. Now the data seemed, on balance, if not absolutely, to support this view. He contacted Svanberg and expressed his wish to stand down.

Perhaps unsurprisingly, given the record of the previous few months, news of the discussions leaked out into the press. On 21 July, *The Times* ran a story headlined: 'Tony Hayward is to step down as BP's chief executive within the next ten weeks'. The story took the UK market by surprise: Hayward's departure from BP had seemed highly probable, but no one had expected an announcement so soon. Ironically, BP's press office had been kept out of the loop by senior management, for fear of such a leak. Consequently, when reporters called to ask if the story was true, they were erroneously told that Hayward's departure was not even under discussion.

But over the following days, a deal was indeed hammered out. Key to the agreement must have been ensuring that it did not affect Hayward's or BP's legal liabilities in America. Experts make the point that a resignation had to be avoided as this would have implied culpability on his part. Whatever went on behind closed doors, the final wording agreed was that Hayward was stepping down 'by mutual agreement'.

On 27 July, 12 days after Macondo was finally capped, Tony Hayward's departure was formally announced. The news was released at the same time as second-quarter results, which included BP's biggest ever quarterly loss due to a $32 billion charge for the expected costs of the oil spill. It was a deeply emotional occasion for Hayward. He had been a BP lifer, someone who, to use his own language, had 'green and yellow blood in their veins', but who, as he saw it, had become 'the public face and was demonised and vilified . . . Life isn't fair.'

In addition to receiving his pension worth over £11 million, and one year's salary of £1 million, Hayward was appointed to the board of TNK-BP, the company's Russian venture. The non-executive role would pay a salary of $150,000 a year and allow him to keep his hand in at the top level of the oil game. He also retained rights to stalled payments under the long-term incentive plan. It can be argued that, had these been axed, it would have played into the hands of litigants against the company by implying he was sacked for bad performance.

The benefit of hindsight

The government investigations into the oil spill might have threatened Hayward's job eventually, but what really did for him in the short term were the gaffes. As he said himself on announcing his resignation: 'BP cannot move on in the US with me as its leader.'

Many argued that his PR handlers should have exercised a firmer hand to prevent him from damaging himself and the company in the ways he did. But the fact that Gowers was not long in his job at the time of the crisis did not put him in a strong position to challenge Hayward's decision to blame Transocean or to take such a public role. How little Hayward was prepared to be guided by his head of press was evident in the way that, on the very day he heard about the blast, he called Gowers' predecessor Roddy Kennedy into his office to advise him on what to do. 'Roddy was Tony's safety blanket,' one insider said. Had Kennedy been co-opted into the response effort for the long haul, he might have been able to convince Hayward to take a lower profile. The fact that, for the two years Kennedy spent as Hayward's head of press, he steered the CEO away from giving any interviews, suggests this certainly would have been his advice.

Brunswick should have pressed harder for Hayward to stand back but, like any service provider, the firm had more incentive to flatter their paymaster than to challenge him. Svanberg was blamed by critics for not getting Hayward back sooner, but he had tried – and his position was not established enough to force Hayward's hand. If he was guilty of being weak in this respect, it was a weakness shared by the whole board, who conversed each week and certainly had an opportunity to make their views known.

Hayward himself did not seek to blame those around him, except perhaps for what he saw as a lack of loyalty on the part of Parker. As he worked towards his September departure, he took to using four-letter expletives to refer to the PR man. Parker himself remarked to a colleague that Hayward had stopped talking to him.

Peer pressure

In the last few months of his tenure at BP, Tony Hayward found himself not only engaged in battle with his own boss and board, but also with his peers at the top of Big Oil.

As a man whom US Senators praised as 'a person of faith', Jim Hackett's first emotion when he heard about *Deepwater Horizon* was doubtless great sorrow for the men who died on the rig. As chief executive of Anadarko Petroleum, 25 per cent owner of the Macondo well, Hackett's next response was one of alarm about the potential financial impact. The feeling was tinged with a sense of relief, however. It could have been worse. Anadarko might have had the misfortune to be the operator of the block. 'Thank God it's you guys,' he told Hayward, 'because we couldn't do this.'

The comment referred to the massive logistical effort unleashed by BP to tackle the spill. But it also encapsulated Hackett's feelings regarding the financial liabilities.

Anadarko was a silent partner in Macondo. It provided 25 per cent of the cash, would receive 25 per cent of the profits, and was liable to pay 25 per cent of the costs should anything go wrong. But the contract between the operating company (BP) and the field investors (Anadarko and Japan's MOEX) envisaged BP paying up front, and then seeking reimbursement from its partners. The question for Hackett was: should he pay up?

Hackett was unusual in the US oil industry as a chief executive with charisma, good looks, style and a willingness to cultivate a high public profile. He also saw himself as a deeply moral man. He had planned to fight for his country as a pilot, before bad eyes and the end of the Vietnam War ended what he called his 'dream'. He gave generously to charity and spoke publicly about his daughter's harrowing sexual abuse and subsequent mental trauma, in an attempt to reduce the stigma surrounding mental health and encourage better treatment.

But Hackett didn't become the best paid executive in Houston, earning over $23 million in 2009,[19] considerably more than Exxon's Rex Tillerson, by spending shareholders' cash like a philanthropist. At the outset of the BP crisis, Anadarko made its position clear. While Hayward was saying BP would not seek to hide behind a $75 million legal cap on its liability for compensation payouts,[20] Anadarko's chief counsel Bobby Reeves told investors that Anadarko anticipated doing just that.[21] And when Hackett spoke to his lawyers, it looked as if the situation might be even better than Reeves had suggested. The lawyers said Anadarko might not have to pay anything at all. While the contract with BP seemed watertight, there was one loophole. If Anadarko could prove that BP had committed 'gross negligence or wilful misconduct', BP would have to foot the clean-up bill, as well as any compensation payouts and fines, alone.

On 18 June, Anadarko issued a statement accusing BP of being 'reckless' and committing 'gross negligence or wilful misconduct'.

Hackett said, 'We were not on the rig, and we were not consulted about the practices and procedures used on the rig floor,' adding that he felt a 'sense of shock and dismay over what we've heard in the public testimony'.[22] He didn't mention that BP had provided Anadarko with daily updates on the drilling operation.

On the same day, Anadarko filed a court action seeking to get out of a contract to hire a rig that it could no longer use thanks to the moratorium on deepwater drilling imposed by President Obama in the wake of the spill.[23] Other companies also found themselves unable to use rigs they had booked. Most of them had agreed to keep hold of the rigs for the time being, rather than pass the cost on to the drilling contractors and push roughnecks out of work. But Hackett was clear on Anadarko's position, no matter how bad it looked: 'When all is said and done,' he said, 'we're going to be protecting Anadarko shareholders.'[24]

Turning on their own

In the week after the rig blast, Tony Hayward got on the phone to the chief executives of the world's biggest oil companies. Exxon's Rex Tillerson, Chevron's John Watson and Shell's Peter Voser all received calls asking for assistance. The CEOs were sympathetic at first and agreed to provide some of their best engineers. (Although not for free – invoices would follow in the post.)

Then, as their men in BP's crisis centre began to feed back information about how BP had drilled Macondo, Tillerson, Voser and Watson's obliging mood shifted. The discovery that BP had relied on a single barrier – the flawed cement at the bottom of the well – to prevent oil from escaping from the reservoir surprised them all. It would be most unusual for a major to drill a deepwater well, let alone a high-pressure one, with a single barrier.[25] Macondo's meltdown may well have been exacerbated by equipment failure, in the case of the BOP, or by human error on the part of the individuals who misread the negative pressure test, but as far as the Big Oil bosses were concerned, BP's drilling practices were primarily to blame for the explosion.

BP now had a third major US disaster in five years to add to Texas City and Alaska, and the Big Oil bosses knew BP would not be the only one to pay for it. Expanded offshore drilling was clearly off the agenda now. Shell knew its controversial plan to drill in Alaska that summer would now not happen, rendering useless preparations it had made at the cost of tens of millions of dollars. New regulations were inevitable. Some politicians began to call for rules that would require oil companies to drill a parallel relief well every time they drilled an exploration well: a prospect that could double drilling costs at a stroke. The CEOs could also anticipate the antipathy towards the industry that the disaster would inevitably generate, as well as the harsher treatment on the taxation front that this might well entail. (Sure enough, in September, a month after the well was capped, Obama said it was time to roll back 'billions of dollars in tax breaks' that the oil companies enjoyed, and in February 2011 he asked Congress to cut $3.6 billion in oil and gas subsidies.)

As the crisis developed, the CEOs of Exxon, Shell, Chevron and Conoco all grew furious at what they saw as their errant peer. In the weeks after the blast, as they conferred about how to react to the disaster, the bosses discussed BP in expletive-peppered terms.

Hayward's comments downplaying the size and environmental impact of the spill added to their rage. The Big Oil bosses knew this would be seen as evidence of a whole industry that gave no care to the damage it caused to the environment. They all knew what had happened when Fred Hartley, president of the Union Oil Company, misjudged public opinion after one of his wells caused a major spill offshore Santa Barbara in 1969. 'I am amazed at the publicity for the loss of a few birds,' he had declared, adding that, since there had been no loss of human life, 'I don't like to call it a disaster.' Hartley's comments had effectively killed offshore drilling in California.

Clearly the whole industry had been negligent in failing to have equipment available to readily cap a blown-out deepwater well, but the Big Oil CEOs felt that this was a moot point since BP appeared to have designed a well that could not be safely capped in any case. They would have to invest in measures to show the industry could tackle any future leak more competently, and with this in mind they began planning a $1 billion response unit. But even this was not enough to save them from being tarred with the same accusation of 'recklessness' that was being levelled at BP by politicians all the way up to the president. It was time for more desperate measures.

The oil industry had long operated a code of *omertà* with regard to safety. Companies did not criticise the work practices of their rivals for fear that this could tarnish them all. Even BP's egregious behaviour at the Texas City refinery had not shifted the industry from this position. But the oil spill was a unique event. The repercussions would potentially be wider than any previous oil industry accident. It required an altogether different industry response. For the first time, the solidarity between those at the top of the oil industry was broken.

In mid-June, almost two months after *Deepwater Horizon* exploded, the bosses of all the Big Oil companies were invited to testify at the House Energy and Commerce Committee hearing. The CEOs lined up at a long table in front of a dais on which two rows of Congressmen were seated. One by one, BP's peers agreed that all the items of concern highlighted by the Congressmen regarding Macondo did indeed reflect a breach of industry norms.

'It's not a well that we would have drilled,' said Marvin Odum, president of Shell's US unit. Sitting beside him, BP America boss Lamar McKay stared down at the table. 'The casing design and the mechanical barriers that were put in place appear to be different than what we would use,' said Chevron CEO John Watson. 'We would not have run a full string.' Rex Tillerson concurred: 'We would not have drilled the well the way they did . . . We would have tested for cement integrity before we circulated the kill-weight mud out.' McKay, with heavy black bags under his eyes, was not invited to defend his company's well.

But the Democratic Congressmen who led the Committee had not invited the CEOs simply to give them a platform to save their own skins. The Congressmen opposed expanded offshore drilling and wanted to show that none of the companies were capable of handling a leak like Macondo. To this end, they quoted from the companies' own oil spill response plans, which in the light of Macondo seemed hopelessly optimistic, not to mention slapdash.

BP, Exxon, ConocoPhillips and Chevron's plans included measures to deal with the impact of an oil spill on walruses, creatures that had not been seen in the Gulf of Mexico for 3 million years. 'It's unfortunate that walruses were included,' the usually stern-faced Tillerson said, fighting back a smile. Two of the plans listed a Dr Lutz as an expert who could provide technical support in the event of a spill. Lutz had died in 2005, four years before the Exxon plan citing him was filed. Tillerson's defence of this bordered on the Haywardesque: 'The fact that Dr Lutz died in 2005 does not mean his work and the importance of his work died with him.' Meanwhile, another plan listed a web link for the Marine Spill Response Corporation – an industry-funded body with a fleet of skimmers and stocks of dispersants – which actually led to a Japanese entertainment website.

The news coverage of the hearing would largely focus on these mistakes and on the general roasting the Committee members gave the executives. But it was still significant in putting on record that all-out hostilities had begun between BP and its rivals; these were positions that only became more refined and entrenched as the months passed.

BP maintained that the rig blast could have happened to anyone, and that it reflected the additional risks being incurred as companies pushed into ever-deeper waters. 'It's been easy for some parties to suggest that this is a problem with BP,' Hayward told a UK parliamentary committee before standing down. 'I emphatically do not believe that that is the case. I don't want to defend the industry because in cases like this it was indefensible.' Nonetheless, BP's smaller UK rival BG Group told the same committee that it appeared 'the Macondo blowout was significantly attributable to a flawed well design', directly contradicting BP's position.[26]

Repeatedly in the months after his appointment, Hayward's successor Bob Dudley said the disaster was a 'wake-up call', not only for BP but for the whole industry.[27] 'Our industry is taking on more and more risk now,' he said in an interview. 'The key for BP and all the companies in the industry is we have got to show that we can manage that risk very, very carefully.' The industry maintained a more clear-cut perspective, however: the accident had happened because BP had been sloppy and had consistently tried to cut costs too far. BP said its peers were simply trying to limit the fallout on themselves from the accident.[28]

It was an argument that was still raging a year after the disaster. In March 2011, at the CERA Week conference in Houston, the biggest event in the US oil calendar, Bob Dudley reiterated BP's case. 'I think it would be a mistake to dismiss our experience of the last year simply as a "Black Swan",' he said, 'a one-in-a-million occurrence that carries no wider application for our industry as a whole.'

But the industry was not won over. 'This conclusion that this is a bigger problem for the industry is just wrong,' said Rex Tillerson when he addressed the conference the next day. 'I think those comments are a great disservice to this industry.'

10

Rebranding an Oil Spill

The oil spill had a devastating impact on the Gulf Coast's environment. Hundreds of endangered sea turtles – including the rarest of them all, Kemp's ridley turtles – were killed by the oil. Dozens of dolphins were killed at the height of the spill, and when the first new calving season began in early 2011, 65 newborn or stillborn dolphins were washed up on the shore. The rate of deaths was 10 to 15 times the normal level, scientists said. Whale sharks and sperm whales – also endangered species – had also been seen swimming in the oil, and were likely affected too.

Thousands of birds were covered in oil, including endangered pelicans. Photographs of a large brown pelican, Louisiana's state bird, trying desperately to flap its enormous oil-drenched wings became the signature image of the spill.

Aside from the deaths and oiling of animals, scientists are beginning to discover worrying signs of potential long-term consequences of the spill, such as toxins in oysters and shrimp. Crab larvae have been born with previously unseen orange globules – believed to be an oil-dispersant mix – in their bodies. Billions of larvae are produced each year and they represent a vital part of the food chain, suggesting the possibility of a wider impact.

Scientists still disagree about the fate of the oil that leaked from Macondo. The White House – ever keen to put behind it a crisis that makes it seem impotent – issued a report in August 2010 claiming 75 per cent of the 4.9 million barrels of spilled oil had been burnt, collected, evaporated or broken down naturally. But the veracity of the report was widely questioned; indeed, the administration was forced into an embarrassing climbdown from its claim that the study had been peer-reviewed. One government scientist even contradicted the report, telling a Congressional hearing that 75 per cent of the oil still remained in the Gulf in late August 2010. A report from the University of Georgia suggested 80 per cent of the oil remained in the water.[1]

What was not in dispute was that, even assuming the rosiest scenario, the damage would be long-term. 'The common view of most of the scientists inside and outside government is that the effects of this spill will likely linger for decades,' said Dr Jane Lubchenco, head of the National Oceanic and Atmospheric Administration, presenting the government report in August 2010. This was hardly surprising given that oil was still being found just below the surface of the shore on Prince William Sound 22 years after *Exxon Valdez*. Certainly, a year on from the BP spill, weathered crude was still being washed ashore and marshlands were still covered in oil.

The exact damage done to the Gulf will probably never be known. Many affected areas, such as the deepwater, had been little studied previously, so scientists can't be sure what the habitats originally looked like. The absence of such 'baseline' data and a shortage of funds for research have made it hard to gather evidence of damage that would stand up in court – a requirement in making BP pay for environmental repair.

The human impact of the spill was also severe. At the height of the crisis, 86,985 square miles, or 36 per cent, of federal waters in the Gulf were closed to fishing. Fifty per cent of Louisiana's oyster beds were destroyed when Governor Bobby Jindal ordered fresh water from the Mississippi be diverted into brackish bays where the beds were located, in an attempt to push back the oil.

Tourism died all along the Gulf Coast – even beyond areas affected by the oil. Unsurprisingly, visitors refused to risk their vacations being ruined by oil-soaked beaches and simply booked holidays elsewhere. The consequent economic uncertainty from all this was blamed for a decrease in mental health in the area.[2] A Gallup survey of nearly 2,600 residents revealed that medical diagnoses of depressive illness had increased by 25 per cent in the three months after the rig explosion. A University of Maryland School of Medicine and University of Florida study published in February 2011 also showed the oil spill 'had significant psychological impact on people living in coastal communities, even in those areas that did not have direct oil exposure'. In May 2010, William Allen Kruse, a charter boat captain whose game fishing business was killed by the spill, took his own life. His family blamed the oil spill.[3]

Domestic violence also rose, with calls to the National Domestic Violence Hotline from Gulf coast states, and in particular Louisiana, spiking between April and June 2010. Local media reported elevated levels of domestic violence, even months after the well was capped, and linked the trend to worries surrounding people's jobs and incomes.[4]

None of this was a surprise to mental health professionals. After the *Valdez* spill, marriage break-ups, alcoholism and suicide spiked, especially among fishing communities. Despite this track record of mental anguish, Ken Feinberg, the man appointed to administer BP's $20 billion Gulf Oil Spill Restoration Fund, said he would not pay out money for mental health impacts.

Doctors also raised concerns about the long-term impacts on people's physical health, especially the thousands involved in the clean-up work, who were exposed to the oil and to dispersants for long periods. Like the environmental damage, it will take years to know how badly affected people were.

Thank you, BP

Despite the signs of lingering damage, BP launched an expensive advertising campaign titled 'Voices from the Gulf' to showcase the recovery of the region. These slick one-minute adverts, run across America and Canada in early 2011, depicted local restaurateurs, fishermen and small tourist business owners on pristine beaches or waters, thanking BP for keeping its promise to 'make it right'.

'BP said they were going to clean it up, help people impacted. They kept their word,' Ike Williams of Ike's Beach Service at Gulf Shores, Alabama was shown saying. He was not compensated for his appearance, the advert said, and nor were the others who contributed to these feel-good ads. The willingness of the victims of the spill to freely thank BP and acknowledge 'mission accomplished' added weight to the growing impression that the spill was a disaster in the past.

But the adverts also highlighted one of the unusual facts surrounding the oil spill: that it was in the victims' interest to downplay the extent of the crime against them. A stigma still hung over Gulf Coast seafood in early 2011, while the tourist trade faced the risk of another lost holiday season. People in the Gulf were desperate to get back to normal, even if this 'normal' was based on overly optimistic assumptions.

Independent scientists were frustrated by overwhelming message of recovery. 'Why can't we scientists produce equally appealing, public service announcements as a reminder that everything is NOT ok in the Gulf,' blogged Dr Holly Bik of the University of New Hampshire, who was involved in the response effort. 'That we won't know the impact of the spill for years. That Gulf food webs (often ending in human consumption) could be concentrating oil-derived toxins. Unfortunately, I don't have millions of dollars at my disposal.'

As it happened, the campaign wasn't entirely successful, and a year on from the blast businesses were reporting that trade remained down because of lingering worries about water quality. Mississippi charter boat operators said bookings had fallen 50 per cent compared to pre-spill levels.[5]

The bill

On a broader level than the economic hit to the Gulf region, the American taxpayer also came out a big loser in the spill. The US government lost its share of the revenues the spilt oil and gas would have generated – over $250 million – but the real hit came from the fact that most of BP's costs in tackling the spill were tax deductible. In total, if BP's estimate of the total bill is correct, the taxpayer will receive $14 billion less over the coming years from BP than it would have done if there had been no spill.[6]

BP's shareholders also face a big economic hit. It is hard to put a figure on the cost of the spill and the company has been forced to go back and raise its estimate a number of times. In April 2011 the estimate hit $41.3 billion. Yet even this enormous prediction was based on two rosy assumptions: first that BP would not be found grossly negligent, and secondly that it would not be forced to pay punitive damages to victims.

If BP is found to be grossly negligent, the potential fine it could face under the US Clean Water Act would rise from around $5 billion to over $21 billion. BP's apparent confidence that it can avoid a finding of gross negligence comes in spite of the fact that the White House has said it expects the higher levels of fines to be imposed. Punitive damages – damages over and above the actual economic loss suffered by victims – could easily add another $20 billion to the bill.[7] Furthermore, if BP is found to have been grossly negligent, its chances of forcing its partners to pay their share of the cost will evaporate.

A year to the day after *Deepwater Horizon* erupted in a fireball, BP filed lawsuits against driller Transocean, cementing specialist Halliburton and Cameron International, maker of the failed blowout preventer, seeking payments that, combined, exceeded $120 billion. BP accused the companies, most notably Transocean, of causing the rig blast through 'fraud', 'misconduct', 'concealment' and 'violations of maritime law'. Despite all the criticism it had received in the Presidential Commission's reports, BP argued its own staff and executives could in no way be held culpable for the accident, and that the company should not have to pay a penny towards the damages.[8]

Moving on

Tony Hayward wasn't the only high-profile departure once the immediate crisis of the oil spill had passed. Andy Inglis, head of exploration and production, and Doug Suttles, chief operating officer of that unit, both stood down in the months after the spill.[9] Inglis took only a few months to find himself another job paying a seven-figure package, with oil services provider Petrofac, where former BP deputy CEO Rodney Chase was chairman.[10]

PR director Andrew Gowers also went. In the months after the well was capped, he raised a few eyebrows, given the perceived failure of the BP PR campaign, by giving speeches about crisis management at PR industry conferences. He had aligned himself closely with Hayward and his man had lost. Chief driller Barbara Yilmaz was reassigned to help with the Gulf Coast restoration effort,[11] and was replaced by a career-long driller. BP's human resources boss, Sally Bott, also left, although for greater things: taking up the same role at Barclays Bank.

In the months after his departure, Tony Hayward continued to downplay the damage to the environment, declaring publicly on television[12] and in a public speech[13] that 'All of the oil is now gone', even as BP employees were still mopping it up. Despite this and his earlier gaffes, he enjoyed a groundswell of sympathy in the UK. The official account of his tenure, offered by his supporters and BP, was that Hayward had made great strides in improving safety but that he simply had not had the time needed to rectify all the shortcomings that had developed under John Browne. Newspaper commentators,[14] UK business leaders and many investors accepted this view – despite the fact that, when it was challenged by the likes of safety regulator OSHA or Congressional committees, BP had failed to present convincing evidence to support it.[15] People quickly forgot that Hayward himself had been in charge of the exploration and production unit for the second half of John Browne's tenure.

Hayward's reputation was strong enough that, when secretive commodities trader Glencore announced it planned the biggest ever flotation on the London Stock Exchange in 2011, it named him senior non-executive director. He almost ended up working under Browne again, as his former boss was shortlisted to be non-executive chairman of the $60-billion Swiss-based trading house.[16] In the end, the group opted for another candidate. Commentators speculated that Glencore's billionaire CEO Ivan Glasenberg may have opted against Browne due to fears he might be too interventionist – an ironic twist of fate given that strategic clashes had been the source of much unease between him and Peter Sutherland.[17]

Friends said Hayward also planned to put together a multibillion-dollar fund to invest in natural resources with Nat Rothschild, scion of the banking dynasty. In a sign that he had learnt little about dealing with the media, he flatly denied the tie-up despite reams of newspaper reports to the contrary, usually based on information from his associates.[18]

His wife Maureen, a former geophysicist with BP, was rumoured to be toying with the idea of writing a book about him to try and address the criticism he had received during the spill, and she emailed family and friends looking for positive anecdotes about him. Some declined, feeling the venture would likely be counterproductive. It is unlikely that she contacted Carl-Henric Svanberg, whose public statements about Hayward in the months after the spill suggested all had not been forgiven or forgotten. A year on from the spill, the best thing he could say about Hayward was that his long service for BP was 'appreciated'.[19]

At the time of writing, the long arm of the US legal system still hangs over Hayward, Inglis, Suttles and others. The US Department of Justice has said it is investigating a range of possible charges against BP executives, including corporate manslaughter. Meanwhile the Securities and Exchange Commission was investigating BP's statements about the low flow rates and its optimistic estimates of the chances of measures like Top Kill, to see if executives may have been guilty of misleading the public markets.[20]

One group that emerged from the spill remarkably unscathed was BP's external PR advisers. Brunswick Group, which took in millions of dollars in fees in the first month of the crisis, continued to supply a large team of people to help BP handle media concerning post-spill litigation. The agency even secured some new work, advising BP on media relations around yet another spat between BP and its oligarch partners in TNK-BP in 2011.

Of course, the other party who didn't suffer from the crisis was the man who ended up running BP.

New face, same spin

About a month into the oil spill, bookies had begun to give odds on Hayward being replaced. At the beginning, the shortest odds were on Iain Conn to succeed him. By the end, no one in, or watching, the industry was surprised when Bob Dudley was named BP's next CEO – the first non-Briton to hold the job.

Given the political storm and the anti-British tone in America, many believed Dudley's nationality swung it for him. This was unfair. He was also the strongest internal candidate in an industry that rarely makes senior external hires, and he had already proven that he could run a large oil company, having led TNK-BP before becoming director for the Americas and Asia. TNK-BP was Russia's third-largest oil company and under Dudley's tenure it had boosted its output and profits and improved its environmental record. The spat between BP and the oligarchs, which saw him flee Russia in 2008, had also given him experience of running an oil company under attack. He had a calm manner that made him unlikely to make the kind of gaffes for which his predecessor had become famous.

Nonetheless, in substance, the BP message changed little after Dudley's appointment. He denied any core flaw in the company's make-up. 'I don't draw the distinction that the difficulties we've had in the Gulf of Mexico is a flaw with the culture of the company,' he told me on the day his appointment was announced – although he was in fact unable to specify what BP's culture was, in spite of two invitations to do so.

Dudley also stuck to Hayward's strategy of shifting blame, although he expressed it in a more sophisticated way. In September 2010, BP published its internal investigation into *Deepwater Horizon*. Its report, compiled by the company's head of safety, Mark Bly, concluded that eight factors had contributed to the well blowout and subsequent blast on the rig. These included a failure of the cement job at the bottom of the well, the failure of the blowout preventer, and the failure of Transocean staff to spot rising pressure in the pipe. The only factor BP – in the form of a relatively low-level employee – was directly responsible for was the incorrect reading of the negative pressure test, and even this blame was shared with Transocean staff. The report prompted a hostile reaction from BP's contractors, lawyers for the victims and some US politicians. 'This report is not BP's mea culpa,' said Democratic Congressman and seasoned oil-industry critic Edward Markey. 'Of their own eight key findings, they only explicitly take responsibility for half of one. BP is happy to slice up blame, as long as they get the smallest piece.'

A couple of weeks later, in his last outing as CEO, Tony Hayward outlined BP's refined position on who was to blame and for what. 'What we have is a lack of rigour and a lack of oversight of contractors,' he told a UK parliamentary committee looking into the implications for drilling in the UK. 'The contractors here were world class and you might have thought they wouldn't have needed that level of oversight, but it was clearly something that was found wanting.'

So BP was accepting a certain amount of blame, but only for having failed to stop others from screwing up. And, of course, it claimed others in the industry had been similarly negligent. BP was only different in having been unlucky enough to suffer the consequences.

Just as BP had professed an epiphany on process safety in the wake of the Texas City disaster, following the MC-252 incident (as the disaster became known in BP-speak) the company began to profess a great awakening on the subject of contractor oversight.

But the penchant for epiphanies was doubly familiar. Here again, BP was claiming an awakening in an area in which it had previously claimed to have had an awakening. BP claimed to have addressed process safety shortcomings after the Grangemouth refinery accidents in 2000, only to claim after Texas City that it had been ignorant of how important process safety was. Similarly, in 2007, Hayward claimed to have recognised the risks attached to BP's use of contractors. He said he had hired additional staff to watch over external service providers.

Less than a week before *Deepwater Horizon* blew up, BP's top management was again quizzed about the issue. An investor at BP's 2010 AGM had expressed concern about the risks of relying on contractors who might not be up to the job. 'The sad part', he had said, 'is that fiscally and legally the ball still sits in your hands.' Carl-Henric Svanberg had smiled and nodded, and answered, 'Absolutely.' He could not agree more. Hayward had also reassured the investor that there was no problem, because 'What we have been doing is building functional capability and skills', with CFO Byron Grote adding that 'we're very cognisant that you can't flip problems over the wall.'

But if BP's latest epiphany seemed like a PR trick, it wasn't the only one. A month after BP's internal investigation was published, Bob Dudley gave a speech at the annual conference of the main British business lobby, the CBI. He spoke mainly about BP's intention not to quit America, and the initial reports from news wires, websites and broadcasters led on these comments. Reuters did not. When I received an embargoed copy of the speech, I was inclined to ignore it, as it appeared to be a rehash of remarks Dudley had previously made in America – except for one part. Before wrapping up, Dudley gave his views on the media coverage of the oil spill. There had been, he said, 'a great rush to judgement by a fair number of observers before the full facts could possibly be known'. He added: 'I watched graphic projections of oil swirling around the gulf, around Florida, across and around Bermuda to England – these appeared authoritative and inevitable. The public fear was everywhere.'

There was a remarkable familiarity to these comments. Here was Dudley, criticising the media for stoking public fears – exactly the accusation he had levelled five months earlier at scientists who had the temerity to challenge the 5,000-barrels-per-day flow-rate estimate being propagated by BP. I wrote a story that included references to his previous 'scaremongering' comments, which BP had obviously decided no one remembered. The story was immediately picked up by our media clients globally, especially in the US, prompting a stinging reaction from BP's critics. Congressman Markey was quick out of the blocks with one of his trademark pithy emails: 'BP is continuing to point the finger at everyone but themselves,' he wrote. 'Since this disaster began, BP has stood for "Blame Passed".'

The company was aggrieved and complained that my 'rather strange story' had led to the rash of unflattering front-page headlines the following day.[21] But the suggestion that these were throwaway remarks was odd, given that they had been carefully scripted. BP wanted to put the spill behind it and had decided that one way to help do this was to encourage a view that the oil spill had been a media crisis rather than a real one.

Doubtless some in the media over-clubbed their coverage at times. But one could just as easily argue that we took an excessively lenient position at other times. After all, the mainstream media, myself included, had treated BP's 5,000-barrels-per-day estimate of the flow rate as credible for weeks after independent scientists challenged the figure. Mainstream news organisations treated President Obama's claim that the spill was 'the worst environmental disaster America has ever faced' with scepticism, while the coverage of the environmental aftermath of the spill consistently saw optimistic voices – such as those arguing that 75 per cent of the oil had vanished – receiving more column inches and air time than negative voices.

Systemic problem?

As it happened, America was not prepared to accept BP's word on the accident. President Obama created a National Commission to launch an inquiry, while the Department of the Interior commissioned another probe and a third was conducted by the Coast Guard in conjunction with the Minerals Management Service, now renamed the Bureau of Ocean Energy Management, Regulation, and Enforcement (BOEMRE) in order to signal a change from the agency's discredited past.

The National Commission was the first to report, issuing its final assessment in January 2011. Its account of the events that led up to the rig blast was similar to BP's. This was unsurprising, since survivor accounts and computer logs meant most of what had happened was undisputed. But the commission's view of where mistakes had been made and who was to blame was noticeably different from BP's.

In short, the National Commission found that the accident was caused by bad management and 'a culture of complacency' regarding risk. While it agreed with BP that blame should be shared between BP and its contractors, the Commission differed sharply on the balance. BP's report had found contractors guilty of seven and a half out of eight causal factors; the Presidential Commission highlighted nine decisions that had increased risk unnecessarily, finding shore-based BP managers (not those on the rig) responsible for six of them. BP's shore-based managers were also 'perhaps' jointly responsible, with Halliburton, for another risk-augmenting decision, while BP's lead representatives on the rig – Donald Vidrine and Bob Kaluza – were deemed primarily responsible for an eighth decision. According to this report, BP was to blame for seven – possibly eight – of the nine decisions.

The Commission noted that the one thing all nine decisions had in common was that they saved time, and therefore saved BP – rather than Transocean, which was on a daily rate – money. Transocean was criticised for failing to train its staff better and for having overly complex procedures for handling blowout situations. Halliburton was criticised for the way it tested the cement. But it was impossible to read the report as anything other than a damning indictment of BP's performance. The co-chairman of the panel, William Reilly, was certainly clear on this. 'The centrally responsible company in the Macondo blowout was the operator, BP,' he told Bloomberg TV. 'Most of the bad decisions, however, were made by BP or with BP's approval and acceptance.'[22]

The report of the Chief Counsel to the Commission, issued a month after the Commission's report, dug deeper into the root causes of the accident and was even more scathing about BP. 'While many technical failures contributed to the blowout,' it read, 'the Chief Counsel's team traces each of them back to an overarching failure of management.'

The Chief Counsel's report cited as evidence BP's system of incentives that prioritised speed of drilling over safety, its failure to abide by its own operating procedures, lack of training in basic but critical areas, and its decision to use a cheaper but riskier well design. It rubbished the claim touted so vociferously by Tony Hayward that the so-called failsafe mechanism, the blowout preventer, could or should have prevented the disaster. 'BOP failures were NOT [the report's own emphasis] the root cause of the blowout . . . Even if the BOP had functioned flawlessly, the rig would have exploded and 11 men would have died.' BP's well design – incorporating rupture discs into the outer well casing and the choice of a long string production casing without a protective outer casing – was also criticised for making it hard to cap the well after the blowout occurred. BP has not responded to the allegations in this report.[23]

In addition to listing a series of decisions that contributed to the accident, the reports highlighted other decisions which did not do so, but which further illustrated BP's tendency to increase risk while cutting costs. They accepted that BP had failed to oversee its contractors, not least in failing to challenge work by Halliburton about which it had recorded concerns. But the claim that the accident was simply a result of inadequate oversight of contractors was utterly slated. The problem was that BP's own drilling operation was declared to be intrinsically flawed. If Hayward, Inglis and Suttles had not already departed it is difficult to see how they could have survived this verdict.

The Presidential panel also took aim at BP's insistence on downplaying the flow rate: 'If BP had devoted a fraction of the resources it expended on the Top Kill to obtaining a more accurate early estimate of the flow rate, it might have better focused its efforts on the containment strategies that were more likely to succeed.' (A year after the disaster, BP was still arguing about the flow rate in what its critics saw as an attempt to limit its liability for fines – the less spillage, the lower the fines. The company claimed that a third less than the official spill estimate of 4.9 million barrels had leaked – despite previously saying it was impossible to put any estimate on the flow rate.)

But for all the criticism, the Commission's reports also had a silver lining for BP: a controversial piece of logic that would allow the company to stick to its main defence. The Commission concluded that, since most oil companies used Transocean and Halliburton to drill and cement their wells, and since these companies had also displayed serious safety flaws, it was likely there was an industry-wide safety problem. 'The immediate causes of the Macondo well blowout can be traced to a series of identifiable mistakes made by BP, Halliburton, and Transocean that reveal such systematic failures in risk management that they place in doubt the safety culture of the entire industry,' it said.

On the face of it, this is undeniable. If Transocean had failed to train staff on *Deepwater Horizon* how to read negative pressure tests, and if the rig's procedures for dealing with a blowout were too complex to be effective, then the same problems were likely to be repeated on the other rigs in the company's fleet. Similarly, if Halliburton had procedures that allowed a cement job to start without sufficient testing having been completed, this could lead to bad cement jobs elsewhere. But the impression created by these comments – not least because of the way BP cited them continually – was that the Commission believed that all oil companies were equally unsafe. This simply was not the case. The Commission never once suggested it had found a single other oil company whose practices were as unsafe as BP's, and the only oil company accused by Commission staff of having systemic safety flaws was BP.

The industry was predictably furious at the implications of this conclusion. 'I do not agree that this is an industry-wide problem,' said Exxon's Rex Tillerson, adding, 'The Commission did not investigate the entire industry.' On this second point, Tillerson was certainly right. And in so far as Commission staff did study other oil companies' practices – those of Exxon, Shell, ConocoPhillips and Chevron – what they found prompted compliments for these companies.

Tillerson and his peers feared that, if BP's view prevailed, the whole industry would be landed with onerous new regulations. In the end, these fears were not realised. A shift in the balance of power in Congress in 2010, from the Democratic Party to the Republican Party, and a rise in oil prices to above $100 a barrel, stalled moves to significantly toughen offshore safety regulations. When the co-chairmen of the Presidential Commission testified at Congressional hearings in early 2011, they received a harsh grilling from Republican leaders, who questioned their assertion of a systemic safety problem in the oil industry.

Even President Obama's taste for regulatory overhaul was curtailed. The renaming of the Minerals Management Service had a cosmetic feel to it, especially when Obama shied away from material change such as splitting off the oil licensing and safety regulation functions into separate government departments, raising the risk that, in time, the agency's focus would shift back from safety to revenue generation. The lack of material change in the regulatory environment was emphasised when, in early 2011, the BOEMRE awarded the first deepwater drilling permit since the moratorium on such wells had been lifted. It was to a project in which BP had the largest shareholding.

But BP wasn't naive enough to see the green light for its return to the Gulf of Mexico as a sign that everything would quickly revert to normal for it in the US. Dudley knew the company would face headwinds in dealing with government agencies for years to come, and so began to scale back BP's reliance on America. He put over $10 billion worth of US assets up for sale, including the Texas City refinery. Meanwhile, he signed exploration deals with China, India and Indonesia. His crowning glory, however, was supposed to be a $16 billion share swap and Arctic exploration venture with Russian state-controlled Rosneft, intended to give BP access to offshore exploration blocks in the Russian Arctic that were believed to contain over 40 billion barrels of oil, and which in the process would make the Kremlin BP's largest shareholder.

The rapid deal-making and the aplomb with which the Russian plan was announced – I received a mysterious call from BP at 5.30 p.m. one Friday afternoon in January 2011, telling me to be at St James's Square with a photographer and TV cameraman three hours later – brought to mind the Browne era. This, perhaps, shouldn't have been surprising. After all, while the faces had changed, BP continued to be run by former Browne turtles. Refining boss Iain Conn, head of new projects Bernard Looney, head of technology Andy Hopwood, head of BP Russia David Peattie, treasurer Dev Sanyal and Bob Dudley himself had all been trained at Browne's knee.

Dudley's deals failed to impress the market, however. Analysts noted that the new Asian ventures offered low margins, while the Rosneft tie-up sparked a dispute with BP's oligarch partners in TNK-BP, who secured a court injunction blocking the transaction.

Wonderful people

Almost a year to the day after the explosion of *Deepwater Horizon*, BP held its 2011 AGM. As usual the location was the ExCeL Centre in London, and as usual it was a colourful affair. Socialist Party representatives distributed copies of their party newspaper, environmental campaigners held aloft signs showing pictures of oil-stained beaches, and a brass band dressed in red and gold tunics, representing striking BP employees from Hull, played loud music.

Shortly after I arrived, things really kicked off when Diane Wilson showed up. A shrimper turned environmental campaigner from Texas, she had become famous after smearing herself in oil at the Congressional hearing at which Tony Hayward was testifying. She was part of a group of around half a dozen people from the Gulf region who had bought BP shares so they could attend and indeed speak at the AGM. This is a common tactic employed by environmental groups: even when it's obvious that their intention is to disrupt proceedings, they rarely have any difficulty in gaining access. But Wilson's decision to, once again, douse herself in what looked like oil before even entering the building prompted security guards to bar the group's entry. The scuffle and refusal of entry made front pages and TV reports on both sides of the Atlantic, causing further embarrassment to BP as its unwelcome anniversary approached.

Inside the meeting hall, events were almost as dramatic – in corporate governance terms, at least. Some newspaper commentators had predicted that Carl-Henric Svanberg would face a revolt, but as it happened it was the doyen of City columnists, Sir Bill Castell, chairman of BP's safety committee, who bore the brunt of shareholder anger. Forty-three per cent of investors either voted against his re-election or abstained, compared to an average of around 3 per cent for directors of FTSE 100 UK companies in 2010.

It would be heartening to think that this represented the start of a heightened vigilance on the part of BP investors towards safety, but it's hard to be optimistic given their poor track record. The fact is that many fund managers had a sense that BP was not as culturally committed to safety as its rivals. 'There was definitely a different flavour to the presentations they made, compared to the likes of Shell and Exxon, in terms of how technical expertise and safety was regarded within the organisation,' one big institutional fund manager told me. But he also admitted that this had not really played into his investment decisions. 'It's extremely difficult to model,' he added, a little defensively.

BP investors' historic disinterest in safety – even though it represents a major financial risk to them – is not atypical. As the head of investor relations for one of the world's five biggest oil companies told me, 'Investors don't see the benefit of health and safety. When you talk about safety they turn off or say, "Yes, but where is the money?"' And likewise, the people who manage the funds that hold most of BP's shares are rewarded on year-to-year performance. They knew instinctively that

BP's safety culture could potentially cause a problem but they also believed BP's hard driving on costs, or its penchant for transformational deals, could boost the company's shares more than those of its rivals in the short term.

At the end of an AGM, directors usually mingle with investors, but, perhaps due to the way the vote went, Castell wasn't in evidence among the crowd. So instead I ended up chatting to retired US Admiral Frank 'Skip' Bowman, who had joined BP's board, and the safety committee, a few months after Macondo was capped. A short, neat, courteous figure with folksy charm (when I enquired how I should address him he said, 'You can call me anything you like up to and including "Grampy" – my wife calls me "Hey you"'), his appointment had been much vaunted by BP, who said his experience as a former head of the US Navy's fleet of nuclear submarines and ships would help strengthen the company's safety culture. Indeed, the safety culture of the US nuclear navy had been cited in the Presidential Commission's report as an exemplar for BP and the oil industry.

Admiral Bowman said he had been sceptical about BP's commitment to safety before joining the board, but had since been convinced of the company's good intentions. This was partly because of the changes Dudley was imposing, which included overhauling bonus schemes to ensure no one had an incentive to put cost-saving before safety, abolishing independent strategic business units in the upstream division, and pledging to bring some outsourced activities back in-house. But what had been at least as important as all this in removing Bowman's scepticism about BP was getting to know the people there. 'There's a sense that there must be a lot of people at BP who get up in the morning and say to themselves, "I wonder how can I do bad today, I wonder what I can do to cause trouble",' he said. 'Well, that's not true; these are wonderful people.'

Afterwards I spoke to refining boss Iain Conn, who was his usual affable self and appeared genuinely contrite following some harsh questioning from investors regarding the oil spill. Svanberg was there, too, flanked by his partner Louise Julian. The chairman was more tense than he had been the last time we met, over a glass of wine at a BP reception, but he was courteous and seemed appropriately humbled by the shareholders' ire. When the event was over, I rode the Tube back into London with CFO Byron Grote, and as we chatted about John Browne's rebuilding of BP in the 1990s, he didn't try to gloss over the failings of the strategy he had himself helped implement. As with most of my conversations with individual BP managers over the years, I found nothing in these men to challenge Bowman's claim that these were essentially decent people. Yet I found this less comforting than Bowman did.

Corporations are essentially amoral. Their purpose is neither to do good nor bad but to provide returns, over the long term, to their investors. This is not a goal that knowingly threatens the welfare of society, since doing harm to others risks attracting financial penalties or even the loss of a company's licence to operate. Indeed, economists going back to Adam Smith have argued that commercial enterprises' amoral pursuit of self-interest even benefits society, by encouraging the most efficient use of the world's resources – the so-called 'invisible hand' effect. Of course, any positive outcomes for society are incidental to the firms: were these outcomes their primary goals, the enterprises would be charities.

This all means that the extent to which a company is a good social citizen is less a function of its managers' moral compasses than of their competence and perspective on what constitutes a good business model. It is this that determines the corporate structure developed by top management – and in particular the incentives given by them to employees. If they develop badly thought-out incentives and enforce them blindly, there will be unintended consequences, something the financial sector has illustrated on a grand scale in recent years.

Since John Browne's remoulding of BP in the 1990s, 'accountability' has been a cornerstone of BP's corporate culture. Managers are held responsible for their commercial decisions, with successes bringing rapid promotions and failure prompting dismissal or demotion. Some industry peers, politicians and former employees have criticised this culture, but Bob Dudley has defended what he describes as a 'business model that drives for performance'.[24] The question is whether he will be the first BP CEO in almost two decades to make the model work.

Notes

1. The Sun King Rises

1 Information taken from BP website; *The History of the British Petroleum Company: Volume 2* by James Bamberg (1994); *British Petroleum and Global Oil 1950–1975: The Challenge of Nationalism* by James Bamberg (2000); and interviews with BP executives.

2 John Browne, *Beyond Business* (2010). Discussing the refusal of the board, led by Sutherland, to consider his proposal of merger talks with Shell in 2005, Browne disparages the decision, saying it reflected 'Why rock the boat?' thinking. Sutherland has stated publicly that such a deal made no sense. In his memoirs, Browne does refer at one point to 'the chairman', saying he 'was keen to move things on very quickly' when it came to Browne leaving the company in 2007 – but even then, he does not use Sutherland's name.

3 'Sun King of the oil industry', by Tobias Buck and David Buchan, *Financial Times*, July 2002; 'Refilling BP's tank', *BusinessWeek*, 23 July 2007; 'New BP boss may bring changes in style, strategy', by Tom Bergin, 2 May 2007.

4 Interview with Lucy Kellaway, *Financial Times*, February 2010.

5 In 2007, BP America CEO Bob Malone told a Congressional committee that 'extreme budget pressures' had hindered pipeline maintenance. 'BP admits budget a factor in Alaska spill', by Chris Baltimore and Robert Campbell, *Reuters*, 16 May 2007.

6 Interview with *Time* magazine, March 1990.

7 Interview in *The Times*, 6 February 2010.

8 'Sun King of the oil industry', by Tobias Buck and David Buchan, *Financial Times*, July 2002.

9 'Sun King of the oil industry', by Tobias Buck and David Buchan, *Financial Times*, July 2002.

10 Anecdote taken from John Browne, *Beyond Business* (2010).

11 Interview with author.

12 Interview with author.

13 *A History of Royal Dutch Shell, Volume 3*, by Keetie Sluyterman (2007).

14 Interview with author.

15 Interview with *Fortune* magazine, 1987.

16 Browne criticises the structure for being overly complex in his memoirs.

17 Interview with author.

18 Detail on the hotel and its history are based on interviews with conference attendees, the hotel's website and from articles including 'Under New Owners, "Club Fed"', by Bruce Buursma, *Chicago Tribune*, November 1990, and 'A US-run resort is the legacy of Charles Keating', by Susan Taylor Martin, *St Petersburg Times*, December 1990.

19 Interview with author.

20 Interview with author.

21 Speech by Andy Inglis, 'The Changing of the Guard', at Rice University, 14 October 2008.

22 Hayward has cited this anecdote on many occasions, including during an address to students at the Graduate School of Business, Stanford in May 2009.

23 Interviews with executives and reports including 'BP chairman is ousted in boardroom struggle', by Simon Beavis, *Guardian*, 26 June 1992.

24 'Sun King of the oil industry', by Tobias Buck and David Buchan, *Financial Times*, July 2002.

25 In his book *Changing Minds: The Art and Science of Changing Our Own and Other People's Minds* (2004), Howard Gardner explains at length how BP became a 'learning company'.

26 Press release from the awarding body, Teleos, December 1999.

27 'Getting Results Through Organization Design', by E. Craig McGee and Kathy Molloy, *Harvard Business Review*, July 2003.

28 'BP's frontier oil exploration strategy could ensure the survival of the company in the future', *Lloyd's List*, 7 October 1989.

29 Anecdote of the Shell approach is cited in Browne's memoirs.

30 Amoco board member interview with author.

31 'The burden of power', editorial by John Browne, *Global Finance* magazine, 1 February 1998.

2. Climate Change Profiteer

1 The 10 billion barrels figure is cited in 'Giant Oilfield in Colombia Is Too Small for Wall Street', by Thomas C. Hayes, *New York Times*, 30 October 1992.

2 'Bullish BP buys Maxus block', *Petroleum Economist*, 23 December 1993.

3 'BP battles to open up Colombian field', *Lloyd's List International*, 28 April 1993.

4 BP press release, 'BP Makes Major New Oil and Gas Finds in Colombia', 27 July 1995.

5 'Colombia Block Has Fewer Gas Reserves Than Expected', *Emerging Markets Report*, 30 October 1996; 'Colombian Pauto, Florena fields seen on stream early 2001', *Reuters*, 5 January 2000; 'BP Amoco sells Venezuelan oil acreage to Repsol-YPF', *Reuters*, 21 July 2000.

6 'Venezuela celebrates success in oil opening', *Energy Economist*, February 1996.

7 'Big oil ready at the starting blocks in Venezuela', by John Paul Rathbone, *Reuters*, 10 July 1996.

8 Interview with author.

9 Information taken from text of speech.

10 Comments drawn from various newspaper interviews and Browne's memoirs.

11 UN press release, July 1998.

12 BP annual reports.

13 Interview with author.

14 'BP's Hayward: "Extraordinary that Coal Plants Still Being Built in US"', *NGI's Daily Gas Price Index*, 25 March 2010.

15 UMW press release, March 2010.

16 'BP Says Coal-Related Comments Taken Out of Context', *NGI's Daily Gas Price Index*, 26 March 2010.

17 Interview with author.

18 'Oil firms to see windfall from EU CO_2 scheme', by Tom Bergin, *Reuters*, 24 May 2005; 'EU to phase in CO_2 auctions for refinery, airlines', *Reuters*, 22 January 2008; 'EU to set easier CO_2 regime for heavy industries', by Paul Taylor, *Reuters*, 20 January 2008.

19 Various sources, including: 'The great global warming swindle', by Ross Clark, *Spectator*, 8 August 2007; 'NHS carbon trading sees millions go up in smoke', by Melissa Kite, *Sunday Telegraph*, 5 November 2006; 'What type of Europe do we want?', by Labour MPs Colin Burgon, Jon Cruddas, Jon Trickett, February 2008.

20 UK government data.

21 'BP, Lucas set venture for solar power systems', *Wall Street Journal*, 18 February 1981, and company statements.

22 Interview with Iain Conn, June 2007.

23 BP press release.

24 BP release, 'BP Forms BP Alternative Energy', 28 November 2005.

25 'Oil firms' renewable investments lag image', by Tom Bergin, *Reuters*, 3 April 2007.

26 A research note entitled 'Premium rating still not justified', issued on 2 February 2006 by Lehman Brothers' Lucy Haskins and Tim Whittaker, highlights the confusion. Referring to an investor presentation by head of the Alternative Energy unit, Vivienne Cox, Lehman said: 'She also reminded the audience that the investment of $8 billion in this area was "over ten years", something perhaps less apparent from the company's current high-profile advertising campaign.'

3. There's No Such Thing as Santa Claus

1 'BP cuts output forecast again, Q3 profit down', by Andrew Callus, *Reuters*, 29 October 2002; BP statement 'BP Files 6-K With SEC And Updates Outlook', 4 September 2002.

2 'Revealed: links that have dubbed BP "Blair Petroleum"', by Joe Murphy, *Evening Standard*, 19 July 2002.

3 BP Amoco 2001 20F regulatory filing.

4 John Browne, *Beyond Business* (2010).

5 'BP boss rejects Blairite markets', by Larry Elliott and Michael White, *Guardian*, 28 January 2005. Browne also discusses the matter in his memoirs.

6 SFA statement, March 2000.

7 David Whitney at *Knight Ridder/Tribune* first reported regulator comments suggesting this in February 2000 in 'Trade Officials Accuse BP Amoco of Manipulating West Coast Oil Prices', but news of the manipulation was broken by reporter Kim Christensen in *The Oregonian* in January 2001 and was later the subject of a congressional investigation.

8 'NYMEX fines BP $2.5 mln for questionable trades', *Reuters*, 15 September 2003.

9 BP regulatory filing, December 2006.

10 The companies' annual reports in which they would be obliged to report these did not do so.

11 2005 annual report.

12 Information taken from Alyeska website.

13 'Oil company official defends probe of industry critic', *Reuters*, 5 November 1991.

14 Author interview with Hamel, Congressional testimony and news reports.

15 'Alyeska Settles Suit Over Effort to Spy on Whistle-Blower', *Wall Street Journal*, 21 December 1993.

16 'Alyeska president steps down after stormy tenure', *Reuters*, 5 February 1993.

17 'Alaska's wobbly pipeline', *Economist*, December 1993.

18 'Labor ruling favors pipeline whistleblower', by Yereth Rosen, *Reuters*, 28 May 1995; 'Alyeska's Internal Problems Persist: Report Cites Whistle-Blower Concerns', by Jim Carlton, *Wall Street Journal*, 23 September 1999.

19 In his memoirs, Browne credits himself with having 'transformed a company,

challenged a sector, and prompted political and business leaders to change'.

20 'Alaska oil spills raise worries ahead of ANWR vote', by Richard Valdmanis, *Reuters*, 26 July 2001; 'Pipeline break leaks oil, saltwater on Alaska Tundra', by Yereth Rosen, *Reuters*, 18 April 2001.

21 'BP Gives Details of Problems at Prudhoe Bay', by Jim Carlton, *Wall Street Journal*, 9 November 2001.

22 *A History of Royal Dutch Shell, Volume* 3, by Keetie Sluyterman (2007).

23 Deposition of Paul Maslin by Brent Coon & Associates, 13 July 2006.

24 Editorial in October 2000 edition of BP in-house magazine *Horizon*, entitled 'Browne: Take Action to put Safety First'.

25 In his book *Changing Minds: The Art and Science of Changing Our Own and Other People's Minds* (2004), Howard Gardner explains at length how BP became a 'learning company'.

26 *The Modern Firm* (2004) by John Roberts – whom Tony Hayward described in 2009 as knowing more about BP than he did – deals with this; former company executives have discussed it with the author.

27 A BP internal review of its 2004 health and safety audits found that: 'It is apparent that a number of [Business Units] are still unaware that they have to monitor their own HSE management system.'

28 Author interview with executives; AQMD statement, 13 March 2003; news reports from same time including 'Calif. air agency sues BP for $319 mln over tanks', *Reuters*, 13 March 2003. According to an AQMD statement of 17 March 2005, BP did not admit wrongdoing but agreed a settlement with AQMD under which it paid $25 million in cash penalties, $6 million in past emissions fees and agreed to provide $30 million to community programmes directed at asthma diagnosis and treatment and to spend $20 million in improvements aimed at reducing emissions at the Carson refinery.

29 Reid's promotion was noted on his LinkedIn webpage, and mentioned in reports including 'BP had a history of problems; Internal inquiries show firm continued to ignore safety, environmental rules', by Abrahm Lustgarten and Ryan Knutson, *Washington Post/ProPublica*, 8 June 2010.

30 Deposition of Manzoni by Brent Coon, 8 September 2006: 'I can't believe I wouldn't have been, but I don't specifically recall.'

31 US Chemical Safety and Hazard Investigation Board Investigation Report, March 2007.

32 Details of injuries from court documents and interviews with victims' lawyers.

33 Quoted in news reports from Reuters and others, and TV reports.

34 BP published the text of Pillari's remarks on its website.

35 'BP doubles corporate ad budget', by Amanda Andrews, *The Times*, 24 December 2005.

36 BP confirmed this policy to the author. First reported on 30 May 2005 in *Advertising*

Age: 'Ad-pull edicts elicit nary a whimper of protest; Editorial vets: Demands by BP, Morgan Stanley threaten pubs' integrity', by Matthew Creamer. Also reported in UK: 'BP sparks row in US over ad demands', by Terry Macalister, *Guardian*, June 2005.

37 Executives told author about the programme and details are posted on BP America website.

38 www.followthemedia.com posting, 25 May 2005.

39 Cynthia Warner, group vice president of health, safety and environment (HSE) and technology for Refining and Marketing, quoted in *BP Magazine*, Issue 3, 2006.

40 'BP identified low morale among refinery workers', by Tom Bergin, *Reuters*, 24 March 2005.

41 Various reports including 'BP apologizes for statements on refinery blast', *Reuters*, 25 May 2005.

42 Pillari in an interview with the *Wall Street Journal*, 'An Oil Giant Faces Questions About a Deadly Blast in Texas', by Chip Cummins and Thaddeus Herrick, 27 July 2005.

43 Management Accountability Project led by BP executive Wilhelm Bonse-Geuking.

44 Interview with Ian Vann, former group vice president, exploration at BP.

45 'BP Texas refinery blast cost seen under $400 mln', by Tom Bergin, *Reuters*, 1 April 2005.

46 BP 2007 Annual Report.

47 All these figures are outlined in 'How BP's oil spill costs could double', by Tom Bergin, *Reuters*, 1 December 2010.

48 This is based on the author's interviews with Browne's associates and advisers. In his memoirs, Browne says, 'My emotional self prevailed over reason. I did not know how to leave.'

49 In his memoirs, Browne extols the virtues of a CEO stepping up to become executive chairman, in the section in which he discusses his retirement. Two close associates of Browne have told the author that he has considered this and others have also reported the fact, including Tom Bower in *The Times*, 20 September 2009: 'The rise and fall of BP boss John Browne'.

50 Hayward later publicly stated his scepticism of the value of mega-mergers: 'BP says oil majors could merge with state oil Cos', by Tom Bergin, *Reuters*, 3 February 2009.

51 'China rebuffs BP bid for big Sinopec stake', by Tom Bergin and Charlie Zhu, *Reuters*, 13 October 2005.

52 'BP chief calls for retirement age to be axed', by Thomas Catan, *Financial Times*, 10 April 2004.

53 'Browne's last big strike', by Sylvia Pfeifer, *Sunday Telegraph*, 9 April 2006; 'What keeps BP's Lord Browne in the driving seat?', by Rupert Steiner, *The Business*, 14 May 2006.

54 The fact that people were lobbying for Browne to stay was reported widely at the time.

The *Sunday Times* reported on 30 July 2006 that 'Sources close to BP say they believe Browne has waged a private campaign to stay on. They point to "overt" signals such as his public statements on ageism, and more "covert" ones. A source said: "A bizarre collection of business people who have nothing to do with BP have rung newspapers and BP non-executives saying that Browne should be allowed to stay on."'

55 This account is based on the author's discussions with BP staff and advisers over the years. The disagreement was also reported in several newspapers including: 'Boardroom ageism puts Browne and Sutherland in the spotlight', by Carola Hoyos, *Financial Times*, 25 July 2007; 'The Friday night bombshell', by Sylvia Pfiefer, 30 July 2006; and 'Browned Off', by Tracey Boles, *Sunday Times*, 30 July 2006.

56 'The true story about Lord Browne – by ex-rent boy lover', by Dennis Rice, *Mail on Sunday*, 6 May 2007.

4. The Perfect Candidate

1 'BP CEO's 2008 exit kicks off race for top job', by Tom Bergin, *Reuters*, 25 July 2006.

2 Address to students at the Graduate School of Business, Stanford in May 2009.

3 'BP CEO Browne to retire earlier than planned', by Mike Elliott, *Reuters*, 12 January 2007.

4 'Changing of the Guard' note from Citigroup, 15 January 2007, and 'BP: With the regime change, what is now required to get BP back on the road to recovery' note from Bernstein, 15 January 2007.

5 BP Annual Reports.

6 Author interviews with company executives. Hayward also said on BBC's *Money Programme* in November 2010 that he had only planned to stay on 'a couple of years' longer than his actual three-year tenure.

7 'BP to simplify management, cut costs', by Tom Bergin, *Reuters*, 11 October 2007.

8 Analyst presentation, February 2007. Author listened in at the time and the recording is available on BP's website.

9 'BP: With the regime change, what is now required to get BP back on the road to recovery', Bernstein, 15 January 2007.

10 'BP revamp copies Exxon', by Tom Bergin and Jonathan Cable, *Reuters*, 10 October 2007.

11 'BP CEO sees "dreadful" Q3, shares down', by Tom Bergin, *Reuters*, 25 September 2007. The *Financial Times* first reported the comment in 'BP results set to be "dreadful"', by Sheila McNulty and Ed Crooks, 24 September 2007.

12 Briefing with reporters, March 2010.

13 BP plc: Dividend Support, by Morgan Stanley.

14 In address to students at Stanford, May 2009.

15 KPMG website: www.kpmg.com/Global/en/WhoWeAre/Performance/Annual Reviews/ PublishingImages/IAR2008/economic_challenges.html#

16 'BP decides against spinning off green energy units', by Tom Bergin, *Reuters*, 29 July 2008.

17 'BP closing Md. solar manufacturing plant; Firm laying off 320 workers; Business moving to China, India and elsewhere', by Steven Mufson, *Washington Post*, 27 March 2010.

18 Strategy Update presentation to analysts, 27 February 2008. Author listened in by phone.

19 Conn speaking at Strategy Update presentation to analysts, 2 March 2010: 'Turnover in senior management in R&M has approached 50 per cent.' Author listened in by phone.

20 Hayward disclosed target at 2010 Strategy Update presentation.

21 Inglis speaking at BP's 2010 Strategy Presentation for investors and analysts in London, 2 March 2010. Author listened to this at the time and has a transcript; voice recording is available on the BP website.

22 BP's head of drilling, Barbara Yilmaz, told a conference in 2007. Reported at www.drillingcontractor.org/drilling_industry_leaders_experts_gather_at_2007_iadc_annual_meeting-444

23 Presentation by John Sieg, BP Group Head of Operations, Houston, April 2010.

24 'Chapter 5: Overarching Failures of Management', report of the Chief Counsel to the National Commission.

25 'Chapter 5: Overarching Failures of Management', report of the Chief Counsel to the National Commission.

26 'Building a global career around a global business', by Linda Hsieh, *Drilling Contractor Magazine*, September/October 2007.

27 A research note from brokerage Evolution Securities said BP had used this term to describe its performance.

28 Interview with *Sunday Times*, 1 November 2009.

29 'BP plc: 9% Yield Prices in Risk – Add into March 3 Strategy Update', Morgan Stanley, 2 March 2009.

30 BP Annual Report.

31 OSHA statements.

32 Research note from brokerage ICAP, 1 December 2009.

33 'BP discovers third leak at Alaska oil pipe, no output hit', *Reuters*, 22 December 2009; BP statements.

34 Hayward told students in an address at the Graduate School of Business, Stanford in May 2009.

5. Macondo

1 IMF report: www.imf.org/external/pubs/ft/wp/2010/wp1033.pdf.

2 Quote from 'Minerals Service Had a Mandate to Produce Results', by Jason DeParle, *New York Times*, 7 August 2010.

3 Italian oil company ENI's magazine *OIL*, June 2010.

4 'Shell under fire over North Sea rig safety', by Tom Bergin, *Reuters*, 23 June 2006.

5 'BP drops "Crazy Horse" as name for giant oilfield', *Reuters*, 20 February 2002.

6 Company presentation to analysts, May 2009.

7 'It's a marathon, not a sprint', research note by Evolution Securities, July 2009.

8 Mark Bly, BP's head of safety, later admitted that, while the data was beamed to shore in real time, it was not monitored onshore for potential safety problems. (Testimony to the National Academy of Engineering and National Research Council committee, which was commissioned by the Department of the Interior to investigate the Macondo disaster, 26 September 2010.)

9 'BP's Macondo fate echoes that of fictional city', by Kristen Hays and Rebekah Kebede, *Reuters*, 9 August 2010.

10 BP internal report into the rig blast, published 8 September 2010.

11 Official data, some published in BP's internal investigation, shows BP's peers did not, by 2010, frequently use the long string well design for exploration wells. A *Wall Street Journal* investigation first drew attention to the discrepancy: 'BP Relied on Cheaper Wells – Analysis Shows Oil Giant Used "Risky" Design More Often Than Most Peers', by Russell Gold and Tom McGinty, 19 June 2010.

12 BP internal accident report; Presidential Commission report.

13 BP internal accident report.

14 BP internal investigation; Presidential Commission report.

15 Presidential Commission's final report, page 96: 'That BP expert determined that certain inputs should be corrected. Calculations with the new inputs showed that a long string could be cemented properly.' BP internal report, page 22: 'Halliburton OptiCem™ cement model review concluded zonal isolation objectives could be met using 9 7/8 in. x 7 in. long string as production casing.'

16 'BP admits failing to use industry risk test at any of its deepwater wells in the US', Rowena Mason, *Daily Telegraph*, 3 July 2010. BP confirmed to the author its failure to compile safety cases in the Gulf of Mexico.

17 Chapter 5, report of the Chief Counsel to the Presidential Commission.

18 Chapter 5, report of the Chief Counsel to the Presidential Commission.

19 Mark Bly, BP's head of safety, later admitted that, while the data was beamed to shore in real time, it was not monitored onshore for potential safety problems. (Testimony to the National Academy of Engineering and National Research Council committee, which was commissioned by the Department of the Interior to investigate the Macondo disaster, 26 September 2010.)

6. PR Playbook

1 Captain Alwin J. Landry's testimony to Marine Board investigation, May 2010.

2 Hayward interview with BBC's *Money Programme*, November 2010.

3 A BP source told author, and the comment was also reported in articles in the *New York Times* in 2010 and in *Fortune* magazine in 2011.

4 The Presidential investigation would rule that BP, through its two representatives on the rig, called the shots on the rig, while Hayward himself later rowed back on the claims in an interview with the author on 28 May 2010. When asked to confirm his earlier statement, he could only offer: 'To a large extent that is absolutely true. On the *Deepwater Horizon* there were, I think, 126 people; two of them were from BP.'

5 Internal BP document released by House Energy and Commerce Committee, June 2010.

6 The House Energy and Commerce Committee, House Subcommittee on Energy and Mineral Resources, House Committee on Energy and Commerce, Subcommittee on Energy and Environment, House Committee on Oversight and Government Reform, Senate Committee on Homeland Security and Governmental Affairs, and Senate Committee on Commerce, Science, and Transportation were among those who announced plans for hearings or investigations.

7 'BP shares drop as oil spill worsens', by Tom Bergin, *Reuters*, 29 April 2010.

8 Comment from Brian Williams, NBC *Nightly News*, 30 April 2010.

9 Comment from Tom Hudson on PBS's *Nightly Business Report* on 29 April 2010. Others made similar comments.

10 Media Guardian ranking of Media industry figures.

11 Various reports including: 'Gulf Spill May Far Exceed Official Estimates', by Richard Harris, National Public Radio; 'Experts: Oil May Be Leaking at Rate of 25,000 Barrels a Day in Gulf', Ian Talley, *Wall Street Journal*, 30 April 2010.

12 'BP Alaska president blasts state business climate', *Reuters*, 4 March 2008.

13 MMS Statement: 'MMS responds to record low oil prices with royalty relief modifications', 12 March 1999.

14 In interviews with ABC, CBS and NBC.

15 Figures released by the House Energy and Commerce Committee.

16 Senator Barbara Boxer used the term in an interview with CNN in May 2010, and Congressman Ed Markey and others made similar comments.

17 The transcript of this presidential press conference appeared on the White House website.

18 'Oil spill hit a BP buying opportunity – analysts', by Tom Bergin, *Reuters*, 30 April 2010.

19 'BP CEO says Top Kill going to plan, 24 hours to go', by Tom Bergin, *Reuters*, 26 May 2010.

20 Interview with BBC's *Horizon* programme, November 2010.

21 The Department of Justice did later announce it was looking at a range of possible criminal charges against BP managers, including manslaughter. 'BP Managers Said to Face US Manslaughter Charges Review', by Justin Blum and Alison Fitzgerald, *Bloomberg*, 29 March 2011; 'BP shares hit by manslaughter report', by Tom Bergin

and Dominic Lau, *Reuters*, 29 March 2011; 'BP says DOJ, SEC probe trading around oil spill', by Tom Bergin and Jeremy Pelofsky, *Reuters*, 27 July 2010.

22 'BP chief's wife tells of oil spill hate campaign', by Alex Spillius, *Daily Telegraph*, 8 June 2010; 'BP chief's wife: We've suffered hate campaign over oil spill', by Mark Blunden, *Evening Standard*, 8 June 2010.

23 Comments to press, 7 June 2010.

24 'BP hires ex-Energy Dept official for US media effort', by Tom Bergin, *Reuters*, 31 May 2010.

7. Capitol Punishment

1 'As the World Burns – How the Senate and the White House missed their best chance to deal with climate change', by Ryan Lizza, *New Yorker* magazine, 11 October 2010.

2 Letter to Secretary of Interior Dick Kempthorne from Inspector General Earl E. Devaney. Widely reported including in 'Review of US oil agency uncovers sex, drugs scandal', *Wall Street Journal Asia*, 12 September 2008.

3 Letter to Secretary of Interior Ken Salazar from acting Inspector General Mary L. Kendall, and Kendall testimony to House Committee on Natural Resources, 26 May 2010.

4 Transcript of House Committee on Government Reform hearing. Available at: www.gpo.gov/fdsys/pkg/CHRG-109hhrg35339/html/CHRG-109hhrg35339.htm.

5 'Local official faulted in oil lease blunder; Mandeville resident oversees Gulf drilling', *Times-Picayune*, 24 September 2006; Statement from Congresswoman Carolyn Maloney in 2007 on promotion of 'official responsible for oil, gas royalty rip-off'; 'Promotion of Oynes at Interior Department questioned', *Platts*, 6 February 2007.

6 'US DOI promotes official at center of oil, gas lease controversy', *Platts Commodity News*, 5 February 2007.

7 Government Accountability Office (GAO) statement from 21 September 2009 and reports including 'US Ending an Oil-Royalty Program That Was Tarnished by Scandal', *New York Times*, 17 September 2009.

8 'US looking at "variable" oil, natgas royalties', *Reuters*, 8 September 2009.

9 Salazar published a list of his achievements since taking office in July 2010, and the only safety measure he referred to was the rather inconsequential 'Finalized rule on subservice safety valves that incorporated but went beyond industry standards on high pressure and high temperature valves.'

10 In *New York Times* report 'Minerals Service Had a Mandate to Produce Results', 7 August 2010: www.nytimes.com/2010/08/08/us/08mms.html

11 Sheryl Morris, 'Oil industry responds negatively to proposed elimination of MMS', *Inside Energy*, 3 April 1995.

12 Presidential Commission report notes that: 'The Interior Department, however, subsequently took that legislative exemption and unilaterally expanded its scope beyond those original legislative terms.'

13 Fox News interview, 12 June 2010.
14 MSNBC interview, 3 June 2010.
15 'BP PR blunders carry high political cost', by Tom Bergin, *Reuters*, 29 June 2010.
16 'BP warned UK of risk in delayed Libya prisoner deal', by Tom Bergin, *Reuters*, 4 September 2009.
17 'The Special New Relationship: Fresh disclosures show quite how keen the British government was to keep in the good books of the Libyan dictator', *Sunday Times*, 6 September 2009; 'How ministers helped to free Megrahi', *Daily Telegraph*, 2 September 2009; 'UK official, BP fuel furor over Lockerbie – Statements feed view that prisoner release, oil interests were tied', *Wall Street Journal*, 7 September 2009.
18 'Brown backs families' fight with Libya for IRA attacks', *New York Times*, 8 September 2009; 'How victims of Gaddafi's arms deal with the IRA were sold out', *Belfast Telegraph*, 8 September 2009.
19 'Britain "sold its soul" to Libya deal', *Daily Telegraph*, 15 September 2009; 'Libya deal ensures PC killer will never be tried in Britain', *Daily Telegraph*, 15 September 2009; ITV documentary: *Real Crime:Yvonne Fletcher*, September 2010.
20 'Macondo update 4: reaction to Senate hearing', UBS analyst note, 12 May 2010.
21 Footage on YouTube and comments available at: www.huffingtonpost.com/robert-reich/why-obama-should-put-bp-u_b_595346.html.
22 Dudley interview on BBC's *Money Programme*, November 2010.
23 BBC's *Money Programme*.
24 'BP's Show of Support May Not Save CEO Hayward', by Stanley Reed and Brian Swint, *Bloomberg News*, 10 June 2010.
25 BBC radio interview, 10 June 2010.
26 'UK government says ready to help BP over spill', *Reuters*, 10 June 2010.
27 Hayward interview on BBC's *Money Programme*.
28 Chairman to investors after 2011 AGM.

8. Subsea Slip-ups

1 Interview with BP vice president Kent Wells, 24 May 2010.
2 Quoted in Presidential Commission report.
3 Quoted in Presidential Commission report.
4 BP internal investigation report.
5 Chapter 4.1, report of the Chief Counsel to the Commission.
6 Author asked BP three times to confirm the data in the report of the Chief Counsel to the Commission and to comment on the apparent under-clubbing, without any response. The analysis was confirmed by author with drilling experts.
7 Exxon CEO Rex Tillerson's testimony to the Congressional Committee on Energy and Commerce hearing on 15 June 2010. Shell confirmed its practice to the author.

Industry experts told the author it is common practice among many oil companies to set the lockdown sleeve at the outset of drilling.

8 Report of the Chief Counsel to the Presidential Commission: 'BP's decision to install rupture disks at Macondo and not to use a protective casing complicated its containment efforts and may have delayed the ultimate capping of the well.'

9 'Memo reveals "fundamental mistake"', *Upstream*, 28 May 2010.

10 'We know the pressure in the reservoir is somewhere around 9,000 psi', 3 June 2010.

11 Interview on BBC *Horizon*'s *Deepwater Disaster: The Untold Story*, broadcast November 2010.

12 On the BBC's *Deepwater Disaster: The Untold Story*, Pat Campbell of Wild Well estimated that the decision to abandon well-capping measures unnecessarily prolonged the spill by about three weeks. BP declined to comment on Campbell's allegation.

13 Presidential Report: 'The understated estimates of the amount of oil spilling from the Macondo well appear to have impeded planning for and analysis of source-control efforts like the cofferdam and especially the Top Kill.'

14 As head of the operation, Thad Allen had to approve all the measures.

15 Interview for the BBC *Horizon* programme.

16 At a press conference on 14 July 2010, Thad Allen said: 'But the maximum pressure we were able to achieve pumping the mud down the well bore was only around 6,000 psi.'

17 'Oil spill team eye long fight against "the blob"', by Tom Bergin, *Reuters*, 7 June 2010.

9. Corporate Cannibalism

1 Sources told author and the fact was reported, including in 'Niall Fitzgerald favourite to take over chair of BP', *Independent on Sunday*, 22 February 2009.

2 'BP favours Rio Tinto's Skinner as chairman', by Ed Crooks, Rebecca Bream and Kate Burgess, *Financial Times*, 28 October 2008.

3 'Rio Chairman pulls out of race for BP job', by Tom Bergin, *Reuters*, 16 February 2009; 'BP quandary as doubts rise over Skinner pedigree', by Ed Crooks and Kate Burgess, *Financial Times*, 31 January 2009; 'Rio's chairman reels from wave of investor anger at Chinalco deal', *The Times*, 1 February 2009; 'Rio Tinto revolt halts chairman's move to BP', *Observer*, 15 February 2009.

4 'Ericsson CEO joins BP with strong record, one glitch', by Tom Bergin, *Reuters*, 25 June 2009.

5 Briefing for reporters including the author, 27 April 2010.

6 Speaking at BP's 2011 AGM, attended by the author, Svanberg made it clear that the decision was Hayward's. 'Tony himself was very clear that he wanted to make sure that in America he was seen as Number 1 . . . There was a concern from Tony's side that if I had a more visible role, his role would be diminished.'

7 'Tata-BP opens new line', *Religare Technova News Service*, 29 April 2010.

8 'Castell faces tough questions at BP', City Spy, *Evening Standard*, 4 June 2010.

9 'Disappearing act', People Column, *Financial Times*, 8 June 2010; 'BP: Gloom at the top', by Ed Crooks and Kate Burgess, *Financial Times*, 23 July 2010; 'There will be BLOOD; A "new BP" could grow out of the gulf oil spill – safer but short on assets', *Sunday Times*, 20 June 2010.

10 Sources told the author. The *Daily Telegraph* reported on 11 June 2010 that: 'There has been a changing of the guard at BP, with Tony Hayward, the oil giant's embattled chief executive, sent home from the US . . . Board reshuffle sees Hayward sent home . . . Meanwhile, Carl-Henric Svanberg, BP's low-profile chairman, is to take on a more active role', without citing its source.

11 'Svanberg links BP's future to response', by Ed Crooks, 26 May 2010.

12 *Dagens Industri*, picked up by Reuters: 'BP chairman: board has not discussed CEO's future', 17 June 2010.

13 'The Safest Job In The World', *Newsweek*, 4 June 2010.

14 Letter to Hayward from Committee Chairman Henry Waxman and Congressman Bart Stupak, 14 June 2010, released by the Committee.

15 Chris Blackhurst, *Evening Standard*, 21 June 2010.

16 BBC's *Money Programme*, November 2010.

17 'Chief moves to assure staff that worst is over', by Carola Hoyos, published in the *Financial Times* on 24 June 2010, gave a colourful account of one of these.

18 'BP's Hayward launches City charm offensive: Chief tells investors he will not be stepping down', by Terry Macalister, Julia Finch and Jill Treanor, *Guardian*, 26 June 2010.

19 'Anadarko's CEO at the top of Houston pay list 2009', *Houston Chronicle*, 9 August 2010.

20 Interview with author, 30 April 2010.

21 Anadarko conference call for analysts and reporters after the company announced its first-quarter earnings, 4 May 2010.

22 'BP actions before blowout were "reckless" – Anadarko,' by Kristen Hays and Braden Reddall, *Reuters*, 18 June 2010.

23 'Anadarko seeks to cancel Diamond rig contract', *Reuters*, 15 September 2010; 'Anadarko Asks Court to Cancel Drill Rig Contract Over US Ban', *Bloomberg*, 21 June 2010; and others.

24 Anadarko analysts' call, 4 May 2010.

25 This is acknowledged industry practice and also features in testimony of Dr F. E. Beck, Texas A&M University, to the US Senate Committee on Energy and Natural Resources, 11 May 2010, in which he said: 'In the drilling business it is standard practice to always have multiple barriers in place in the wellbore at any given time.'

26 Hayward told the Parliamentary Committee on 15 September 2010: 'I think the [internal BP] investigation makes it pretty clear that this was not an issue of the well design.'

27 'BP gets "wake-up call" and $32 bln in spill charges', by Tom Bergin and Kristen Hays, *Reuters*, 27 July 2010; he also used the phrase on other occasions, including in a BBC interview in September 2010.

28 When Hayward was asked by the Commons Committee why his rivals had turned on him, he answered: 'I think it's perhaps an understandable response given what was going on in the United States.'

10. Rebranding an Oil Spill

1 'Gulf looks cleaner, but trouble lurks below: 2 groups paint a dark scenario about the amount of oil in the sea', by Seth Borenstein, *Associated Press*, 18 August 2010.

2 'Mental health a growing concern after Gulf spill', by Matthew Bigg, *Reuters*; Statement from government's 'Substance Abuse and Mental Health Services Administration: Coping with the spill', July 2010.

3 'Gulf oil spill's mental toll takes a solemn turn', *Miami Herald*, 28 June 2010; and elsewhere.

4 'Rise in domestic violence may be linked to oil spill, claims frustrations, officials say', by David Ferrara, *Alabama Press-Register*; 'Bayou Mayor: Domestic violence "up 320%" since oil spill', BBC News website, 15 June 2010.

5 'A year on, Gulf still grapples with BP oil spill', by Anna Driver and Matthew Bigg, *Reuters*, 15 April 2011.

6 'BP oil spill to cost US taxpayer almost $10 billion', by Tom Bergin, *Reuters*, 27 July 2010. The total rose to $14 billion as the estimated cost of the spill increased.

7 'How BP's oil spill costs could double', by Tom Bergin, *Reuters*, 2 December 2010.

8 'Lawsuits fly in BP's Gulf spill blame game', by Tom Bergin and Moira Herbst, *Reuters*, 21 April 2011.

9 'BP ousts exploration chief, vows to boost safety', by Tom Bergin, *Reuters*, 29 September 2010; 'BP says exploration and production COO to step down', *Reuters*, 11 January 2011.

10 'Former BP drilling boss joins Petrofac', by Tom Bergin, *Reuters*, 5 January 2011.

11 'BP's Gulf Of Mexico Head Survives Reshuffle – Email', *Dow Jones*, 18 October 2010.

12 BBC's *Money Programme*, November 2010.

13 Speech to the Cambridge Union, November 2010, reported in 'How BP's oil spill costs could double,' by Tom Bergin, *Reuters*, 2 December 2010.

14 'Departing BP chief was on the right track', *The Times*, 30 September 2010; 'Tony Hayward: The fall guy', *Petroleum Economist*, 29 July 2010.

15 'Macondo update 4: reaction to Senate hearing', UBS analyst note, 12 May 2010.

16 Sources close to the matter told the author, and it was reported widely, including by the BBC; 'Former BP boss Browne on Glencore shortlist – source', by Clara Ferreira-Marques, *Reuters*, 14 April 2011.

17 BBC Business Editor Robert Peston, who prematurely reported that Browne had already been selected for the job, said on the BBC website that 'My sense is that Lord Browne was more of a stickler for detail than this entrepreneurial company felt comfortable with' on 14 April 2011.

18 Many reports, including 'Glencore role a comeback for vilified ex-BP boss', by Quentin Webb and Douwe Miedema, 15 April 2010; 'Ex-BP Boss Denies Investment Fund Reports', *Reuters*, 6 April 2010.

19 Comments made at 2011 AGM.

20 'BP Managers Said to Face US Manslaughter Charges Review', by Justin Blum and Alison Fitzgerald, *Bloomberg*, 29 March 2011; 'BP shares hit by manslaughter report', by Tom Bergin and Dominic Lau, *Reuters*, 29 March 2011; 'BP says DOJ, SEC probe trading around oil spill', by Tom Bergin and Jeremy Pelofsky, *Reuters*, 27 July 2010.

21 'Dudley hits out at Gulf spill fear-mongering', by Brian Groom, *Financial Times*; 'BP's Dudley slams "rush to judgment"', by James Herron and Guy Chazan, *Wall Street Journal Europe*, 26 October 2010.

22 Broadcast on 6 January 2011. Video available at: www.bloomberg.com/news/2011-01-06/bp-made-most-of-the-bad-decisions-at-doomed-well-reilly-says.html.

23 BP responded to the report but not the allegations. It said it had 'made every effort to understand the causes of the *Deepwater Horizon* accident to help prevent similar events from occurring in the future', and that the Presidential Commission's findings, 'particularly that the accident was the result of multiple causes, involving multiple parties – are largely consistent with those contained in the BP internal investigation report.'

24 Interview with BBC's *Money Programme*, November 2010.

Bibliography

In addition to the interviews conducted for this book and my own experience working in and reporting on the energy industry since the mid-1990s, I have principally relied on official documents such as company announcements, UK and US government reports, transcripts of Congressional Committee hearings and internal BP documents released as part of litigation or government probes.

In writing about *Deepwater Horizon*, I found especially useful the two reports produced by the National Commission on the BP *Deepwater Horizon* Oil Spill and Offshore Drilling (the Presidential Commission), transcripts of testimonies to the Joint US Coast Guard/Bureau of Ocean Energy Management Investigation into the rig blast, and information revealed through Congressional investigations. Reuters' unrivalled bureau network has provided a comprehensive on-the-ground record of BP's activities around the globe, while specialist publications such as *Upstream* and *Lloyd's List* provided access to the nooks and crannies of the industry that the mainstream media cannot reach. The *Wall Street Journal* has produced many wonderfully researched articles on BP over the years and did some of the best reporting on the background to the *Deepwater Horizon* explosion. The archives of countless other publications, including the *Financial Times*, *New York Times*, *Sunday Times* and *Fortune*, were mines of information and I have aimed to give them due credit in the notes.

Also invaluable were research reports produced by banks and brokerages including Morgan Stanley, Deutsche Bank, Bernstein, Goldman Sachs, Barclays Capital, Evolution Securities, Merrill Lynch, Credit Suisse, UBS and Citigroup, and I consulted over 450 on BP alone.

In addition, the following books were particularly useful:

Bamberg, James, *British Petroleum and Global Oil 1950-1975: The Challenge of Nationalism* (Cambridge University Press, 2000)

Bamberg, James, *The History of the British Petroleum Company: Volume 2* (Cambridge University Press, 1994)

Bower, Tom, *The Squeeze: Oil, Money and Greed in the 21st Century* (HarperPress, 2009)

Browne, John, *Beyond Business: An Inspirational Memoir from a Visionary Leader* (Weidenfeld & Nicolson, 2010)

Jonker, Joost, Jan Luiten van Zanden, Stephen Howarth and Keetie Sluyterman, *A History of Royal Dutch Shell* (Oxford University Press, 2007)

Keeble, John, *Out of the Channel: The Exxon Valdez Oil Spill in Prince William Sound* (University of Washington Press, 1999)

Roberts, John, *The Modern Firm: Organizational Design for Performance and Growth* (Oxford University Press, 2004)

Acknowledgements

Writing about what goes on inside a big European company is always a struggle. European executives are less inclined to talk to reporters than their US counterparts, perhaps because Americans are more open in personality or because companies have less of a need to court public opinion, since Europeans rarely own shares directly. European political and legal structures, which demand less transparency, are also likely a factor.

I wrote this book because I felt that, with my background, I was one of only a handful of people who could. While BP's London press team helped with a number of factual enquiries, the company declined to make executives available for interview. When it came to the particularly sensitive matter of the Gulf of Mexico oil spill, over which there is ongoing litigation, even securing confirmation of facts cited in government reports proved challenging. Fortunately, having written several hundred articles about BP over the past decade, I have built up a significant base of contacts, which I was able to tap for this book. Over 100 past and current BP employees and contractors, from rig hands to board members, have contributed to this book. An even larger number of executives from other companies, as well as industry experts, financial analysts and investors, also provided accounts of their involvement with BP over the years. I am extremely grateful for assistance from: Lord David Simon, Sir Mark Moody-Stuart, Tom Hamilton, Greg Bourne, Nick Butler, Keith Myers, Ian Vann, Andrés Peñate, Colin Maclean, Paul Jennings, Oberon Houston, David Bamford, David Weaver, Ross Macfarlane, John Herrlin, Stephen O'Sullivan, Congressman Michael Burgess, former Congressman Bart Stupak, Brent Coon, Karlene Roberts, Hans Olav Bjornenak, Carl Mortished and everyone else who contributed.

This story would never have been committed to paper without my agent, Maggie Hanbury, and the team at Random House Books, including Nigel Wilcockson and my tireless editor, Silvia Crompton.

I must also thank my colleagues at Reuters, with whom I have reported on BP over the years. These include Paul Hoskins, Sarah Young, Rosalba O'Brien, Dmitry Zhdannikov, Melissa Akin, Barbara Lewis, Kristen Hays, Chris Baltimore, Anna Driver, Ed Stoddard, Matt Daily, Eileen O'Grady, Mark Potter, Alex Smith, Doug Busvine, Dan Lalor, Quentin Webb, Andrew Callus and many others. The support of my editors, in particular Chris Wickham, but also Shankar Sitaraman, Simon Robinson, Martin Howell and Sarah Edwards, has been invaluable. Reuters' willingness to afford me leave to complete the book was also much appreciated, but it goes without saying that any personal views contained in this book are mine alone and do not reflect a position on the part of Reuters.

My greatest support has always come from my family, including my father, sister and brothers. If the input of my two young sons has not always been of practical benefit, their tottering or crawling into my room has always been an enjoyable distraction. But more than anyone, I must thank my wife, Sophy, whose encouragement was the single most important element in creating this book.

Index